Science in a Social Context

Galileo and
Copernican Astronomy

A Scientific World
View Defined

Clive Morphet
Newcastle-upon-Tyne Polytechnic

Butterworths

LONDON - BOSTON
Sydney - Wellington - Durban - Toronto

The Butterworth Group

United Kingdom London	Butterworth & Co (Publishers) Ltd 88 Kingsway, WC2B 6AB
Australia Sydney	Butterworths Pty Ltd 586 Pacific Highway, Chatswood, NSW 2067 Also at Melbourne, Brisbane, Adelaide and Perth
South Africa Durban	Butterworth & Co (South Africa) (Pty) Ltd 152—154 Gale Street
New Zealand Wellington	Butterworths of New Zealand Ltd 26—28 Waring Taylor Street, 1
Canada Toronto	Butterworth & Co (Canada) Ltd 2265 Midland Avenue, Scarborough, Ontario, M1P 4S1
USA Boston	Butterworth (Publishers) Inc 19 Cummings Park, Woburn, Mass. 01801

© SISCON 1977
First published 1977
ISBN 0 408 71303 8

Library of Congress Cataloging in Publication Data

Morphet, C   S
  Galileo and Copernican astronomy.

  Bibliography: p.
  1. Galilei, Galileo, 1564—1642. 2. Copernicus,
Nicolaus, 1473—1543. 3. Astronomy—Philosophy.
I. Title.
QB36.G2M67   1977   501   76—57206
ISBN 0-408-71303-8

Typeset by Scribe Design, Chatham, Kent
Printed in England by Chapel River Press   , Andover, Hants.

Preface

This is not a course in the history of science, nor is it even a course on the history of the Copernican revolution although in parts it will closely resemble a historical account. It is, rather, a course intended to explore the nature of scientific ideas, and the way in which they interact with society by focusing on the struggle that Galileo Galilei had with the church over the Copernican theory of the universe. We choose this particular episode because it serves a double purpose. It can illustrate the complex manner in which scientific theories come to be changed but more importantly, because it involved Galileo in a precise statement of his views on the nature of scientific knowledge, we can study that statement (Galileo's *Letter to the Grand Duchess Christina, see* p. 36) and use it to form our own views on the problem.

We are interested in gaining an (albeit introductory) insight into two major questions:
1. What is the nature and meaning of scientific knowledge?
2. How does science relate to its external environment?

Our main concern is with understanding what scientific knowledge is, not with the study of the Galileo affair as such.

There are questions and suggestions for further reading inserted at places in the main course notes (Chapters 1—8). They are followed by teaching notes (Chapter 9) which provide additional background material, suggestions for essays and for discussion questions, and a fuller bibliography.

Read Chapters 1—6 carefully and consider the questions posed in the text. A good understanding of this section is essential before proceeding to consider the issues in Chapter 7; the questions are designed to help you. Think about them, discuss them with your fellow students; check your answers with your teacher or by referring to the recommended books. The teaching notes suggest that a little time is given to this background material but focus mainly on the *Letter* and the other extracts in Chapter 7. These you should read very carefully. Most of them are summarized but the summaries should be treated with caution; they are intended to help you understand (*not to replace*) the extracts. Again there are questions inserted at points in this Chapter; consider them carefully and see how they relate to the extracts. To answer some of them you will have to refer to books on the short reading list in the Appendix.

Finally in Chapter 9 you will find the main issues clearly laid out with some discussion about each and you will find references which you can use to pursue further any issue which interests you.

Chapter One
Introduction

At the beginning of the sixteenth century an educated European man would have known the way in which the universe worked; he would have known that the earth was stationary at its center, that the giant sphere of its perimeter rotated once a day and carried with it the fixed stars, and that between the earth and this stellar sphere were the moon, the sun and five planets each of which traced, within the bounds of its own crystal sphere, a complex yet predictable path compounded out of simple circular motions. Such a view of the universe was fully consistent with his observations and with his religion.

By the end of the seventeenth century an educated man would again have known the way in which the universe worked; he would have then known that the earth rotated on its axis and, like other planets, revolved around the sun on an elliptical path prescribed by the force of gravity.

In the time that had elapsed a revolution had occurred: the Copernican Revolution, named after its initiator. It might be considered to have been bounded in time by the publication in 1543 of Copernicus' *De Revolutionibus Orbium Caelestium* and the publication in 1687 of Newton's *Principia*; in the century and a half between these two dates the educated man postulated above would have had a choice between the old and the new astronomies.

The notion that science affords a set of objective criteria upon which such choices can be made is rendered suspect by an examination of this period of revolution. The old astronomy bore with it a set of assumptions which were consistent with man's experience, both religious and physical; to the new astronomy, in implication if not in conception, nothing was sacred. As a consequence, the eventual clash between Galileo and the Church was not, as it is often presented, a clash between enlightened reason in the form of Galileo and oppressive reaction in the form of the Church. Rather, it epitomized the conflict between one set of assumptions and another, both honestly held, which made the resolution of the choice between the old and the new astronomies less dependent on reason and more dependent on taste or even faith.

Thus, in the sixteenth and seventeenth centuries, there was a period of transition during which consensus among astronomers could not be taken for granted. By studying it we can learn something of the nature of theory change in science. But perhaps more importantly, this period saw elements of a modern scientific outlook extend its boundaries into areas of enquiry where observation and measurement had hitherto been less important than philosophical speculation and *a priori* reasoning.

Galileo Galilei was the key figure in the battle to have the new astronomy accepted by the Church. His campaign to reconcile Copernican theory with the Christian religion led to the establishment of the ground rules on which knowledge from that time was to be structured. Galileo defined a program which clearly separated science and faith, and, if only by implication, left no doubt as to which was the senior partner.

The key document in this program is Galileo's *Letter to the Grand Duchess Christina* in which Galileo sets out his own case. In order to set the stage for it, it is necessary firstly to examine the old astronomy, Copernicus' innovation, and Galileo's particular contribution to the course of events.

Chapter Two
Astronomy up to the Time of Copernicus

The regular events which take place in the sky relate to important events which take place on earth — to the seasons and thus to the timing of harvests, to the tides and to the flooding of great rivers, and to the basic rhythm of life itself, governed by the division of time into day and night by regularly varying periods of light and darkness. Of all of nature's regularities the regularity of the sky is perhaps the most obvious, and evidence of the urge to account for it is widespread in the records of past civilizations.

Before the time of the Athenian dominance of European culture in the fourth century BC the accounts which were produced relied on the operations of gods, both for the existence of the phenomena and for their regularity. Around the fourth century BC, the century of Greece's 'Golden Age', complex astronomical phenomena began to be seen not as the direct product of some deity but as the natural compounded product of simple operations repeated in perpetuity — astronomy became mechanistic. The universe began to be seen in a form that was to last for nearly two thousand years, and the fundamental aspect of that form was the two-sphere universe.

The two-sphere universe

In order to account for the motions of the stars the Greeks assumed that the spherical earth was bounded by a distant spherical shell on to which the stars were attached. This shell rotated daily on a diametrical axis one end of which intersected with it at a point which could be viewed a little above the northern horizon; a point very close to the North Star. Such a set of assumptions was consistent with observation, and the reader might now consider just what the corresponding observations would be.

To account for the relation of the two-sphere universe to observation in this fashion is perhaps to cheat a little. But an important point is incidentally made: The two-sphere universe, unlike the observations themselves, is a product of the human imagination and one of its startling powers is its ability to replace a host of observations with a simple model. To some philosophers whom we discuss later the function of science is no more than this.

Question

It is important to understand the two-sphere universe. Can you sketch the movements of stars above the northern horizon if viewed from northern latitudes? Use the two-sphere universe to derive the sketch.

After the stars, the next heavenly body to be treated quantitatively was the sun. The sun's path through the sky could be seen as composed of two motions: a daily motion from east to west as if the sun were carried along by the stellar sphere, and an additional motion in a counter-direction such that the sun made one complete circuit against the stellar sphere in a year. This path of the sun through the sphere of the stars is called the ecliptic; it is inclined with respect to the stellar equator (*Figure 1*) and thus accounts for the seasonal variation in the sun's altitude.

The problem of the planets

Seven so-called planets were known to the Greeks; one of these was the sun and the others were the moon, Mercury, Venus, Mars, Jupiter and Saturn. The motion of the moon presented its own set of problems which need not be our specific concern here. Our specific

Figure 1 The sun in the two-sphere universe
Figure 2 Retrograde motion of a planet, relative to fixed stars on stellar sphere
Figure 3 An epicycle-deferent system

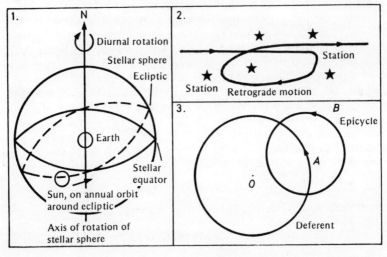

interest is with the five bodies that we now call planets, for the problems posed by the motions of these bodies were solved only through relatively complex developments of the two-sphere universe, and the complexity of these developments was one of the reasons which led Copernicus to re-think the problem.

> ## Question
>
> Redraw *Figure 1* to show the actual path of the sun relative to the earth over the course of a year, i.e. show how the two motions add together.

The motions of the planets observed from Earth are in some respects similar to the motion of the sun: they revolve daily around the earth as if carried by the stellar sphere and they also creep annually backwards relative to the stars. But this backward motion is not as simple and regular as the sun's motion. In its journey around the stellar sphere a planet will sometimes stop, retrace its steps and then continue. Such a looped pattern of behavior is termed retrogression (*Figure 2*).

Aristotle and the crystalline spheres

The cosmology of Aristotle in the fourth century BC had not offered a solution to the problem of the *motion* of the planets, but in this cosmology the planets were recognized and given a place. Aristotle's universe was a universe of concentric spheres, nesting one within the other based on the two-sphere universe but incorporating between the stellar sphere and the earth a crystalline sphere for each of the seven known planets. This spherical construction was reflected in the earth whose four elements earth, water, air and fire were naturally located themselves in concentric spheres with earth central, but disturbed and dislocated by the process of change and decay which took place in the realm of man, below the sphere of the moon. Above the sphere of the moon was perfection, represented by the unchanging aspect of the fixed stars and the unchanging though complex behavior of the planets in their spheres. The Gods were located in a bounding shell, the *'primum mobile'* or 'prime-mover' whose motion was transmitted down through the planetary spheres.

Ptolemy, epicycles, equants and eccentrics

A semi-quantitative approach to the problem of the planets' motion was offered by Eudoxus, a pupil of Aristotle, but the fully quanti-tative solution which lasted for nearly a millenium and a half was developed over several centuries and gained maturity in the work of the Alexandrian astronomer Ptolemy in the second century AD. Ptolemy's solution built on the Aristotelian universe, and on the philosophy of Plato which admitted only uniform circular motion to the realm of perfection, the area of the universe above the lunar sphere.

Question

What was Eudoxus' system? Use one of the recommended texts to find this out.

Plato's insistence on circularity was based largely on *a priori* reasoning. Circular motion is the only motion *perfect* enough to belong to the realm of God's untouched creation; it is 'most perfect and most like to itself', it can go on indefinitely (we would say that it is a cyclic motion, but only the simplest of many possible cyclic motions). Aristotle had taken up these ideas so that in his universe motion in the super-lunary was naturally circular; below the sphere of the moon it was, as we shall see later, naturally radial to or from the center of the earth.

The problem was thus defined by Aristotle and Plato and its solution was a geometric one. Ptolemy showed that simple circular motions could be compounded to correspond to the observed motions of the planets. In order to do this he used three devices; the first is used to account for the gross irregularity of retrograde motion and involves the use of an epicycle on a deferent.

If point *A* describes a uniform circular motion around *O*, and point *B* describes a uniform circular motion about *A* (*Figure 3*) the resulting motion of *B* with respect to *O* will, given certain conditions with respect to the radii and periods of the circular motions, be looped.

The circle centered on *O* is termed a deferent; the circle centered on *A*, an epicycle. If the angular velocity of *B* about *A* is small the motion of *B* about *O* will not be looped but will be a distorted circle.

The second device employed by Ptolemy was the eccentric which placed the center of a planets deferent not at the earth but at some point removed from it. The use of the eccentric is in fact a shorthand device, for the equivalent of motion around an eccentric can be produced by a single epicycle of period equal to that of the deferent; the use of the eccentric is in this sense consistent with Plato's ground

9

rule of uniform circular motion. For the third Ptolemaic device, the equant, this is less clearly the case. Copernicus felt that the equant violated the principle of uniform circular motion and this attitude was doubtless at least partly responsible for the changes that he made. The equant concerns the rate of rotation of a point around the circumference of a circle. Ptolemy defined an equant point which was not the center of the circle, but around which the angular velocity of the encircling point was constant (*Figure 4*). The speed of the encircling point thus varied as it traveled around the circle, and on this point could be built as needed, a further construction of epicycles.

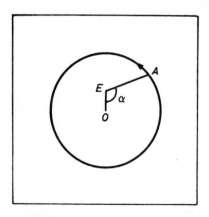

Figure 4　The point A travels around circle center O, but its angular velocity($d\alpha/dt$) is constant with respect to point E, the equant point

By building a system of epicycles on epicycles, incorporating eccentrics and equants, Ptolemy was able to construct an account which used in all about 80 devices and represented the motions of the planets not just approximately but to within the degree of accuracy of the observational data of the time.

It might be noted that there is indeed no limit to the potential accuracy of such a system, for closer and closer approximations to any desired path could always be geometrically achieved by the incorporation of additonal wheels on wheels.

The geometry of Ptolemy fitted into the crystal spheres of the Aristotelian universe to produce an account of the heavens which was not to change, had no need to change, until the Renaissance. In the meantime it was incorporated into Christian theology by St. Thomas Aquinas in the thirteenth century AD and by the sixteenth century was so integrated into the ideology of the times that to interfere with astronomy was to interfere with God's throne.

Chapter Three
Copernicus' Innovation

Copernicus was born in Torun, in what is now Poland, in 1473. He studied at Krakow University and then spent 10 years in Italy at the Universities of Bologna, Padua and eventually Ferrara where he took a degree in Canon Law. He appears to have achieved no great distinction in his studies and eventually returned to Poland where his uncle, Bishop Lucas, had secured an easy living for him as a Canon of Frauenberg Cathedral.

A heliocentric tradition

Copernicus was not an observational astronomer, he made few observations, and those on rudimentary instruments when he could have afforded better. His contribution to the history of astronomy was as a mathematician, or rather as a geometer, for the task that Copernicus set himself was to do a better job than Ptolemy in setting up a geometric account of the motions of the heaven. In particular, Copernicus felt distaste for the equant device which he believed violated the Platonic imperative of uniform circular motion. In producing a geometric account which he felt was superior to Ptolemy's, Copernicus found it expedient to move the earth from the center of the universe and set it in motion with an annual orbit around the sun and a rotation on its own axis.

The geocentric cosmology based on the two-sphere universe had come to dominate Greek thought and had matured into the quantitative astronomy of Ptolemy. However, up to the third century BC a heliocentric tradition of cosmology had existed alongside the geocentric. Heraclides in the fourth century BC had offered a system in which the earth rotated and the planets Mercury and Venus circled not the earth but the sun. Aristarchus in the third century BC offered an account which was in outline identical with the Copernican system which appeared so much later. However, Greek thought had selected the two-sphere universe and its Aristotelian derivative as the model on which to build a detailed quantitative account. As a consequence of the substantial Ptolemaic achievement an alternative approach was no longer felt to be needed and the heliocentric tradition was discarded.

Circles on paper or wheels in the sky?

There is a subtle difference between what might be called a cosmology and what might be called technical astronomy*. The former is concerned to produce an 'explanation' of the universe, not quantitative but in the sort of metaphysical aspect which determines Man's place in the scheme of things. The latter is concerned with the geometry of the skies and technical accuracy is of more consequence than any notion of reality or meaning.

While the distinction can never be absolutely clear, it is perhaps helpful to see Copernicus' innovation as an innovation in technical astronomy. As such it was not intended to be incompatible with Aristotelian natural philosophy, although when taken up by Galileo it proved to be so. It is not truly clear from Copernicus' writing whether he intended the motion of the earth to be a device confined to paper which usefully served as a computation tool, or whether he really saw the earth as a planet in motion through space.

A preface to *De Revolutionibus* was inserted anonymously by the theologian Osiander who supervised the book's production. This preface included the words: 'For these hypotheses need not be true or even probable; if they provide a calculus consistent with observations, that alone is sufficient'†. Copernicus died in 1543, the year that the book was published, and his own attitude to Osiander's preface is not documented. From the text of *De Revolutionibus* and from the fact that Copernicus had delayed its publication for over 30 years it appears that this was a question that Copernicus preferred to avoid. But it was a question that Galileo subsequently attacked with enthusiasm.

The Copernican system

In the Copernican system (*Figure 5*) the sun is at the center of the universe, and around this revolves the planets Mercury, Venus, Earth, Jupiter and Saturn. The moon revolves around the earth. The actual orbital motions of the planets are compounded with epicycles and eccentrics as in the Ptolemaic system; only the equant is missing. The Copernican system used a total of 48 devices, a number comparable with at least some versions of the Ptolemaic, which although when first generated by Ptolemy had incorporated 80 devices had been simplified over the centuries. Moreover the Copernican system was no more accurate than the system that it vied to replace, and it did not at that time explain any phenomena which the Ptolemaic was unable to explain. Its sole technical attraction was that it accounted more basically and more elegantly for certain phenomena which in the

---

*See Chapter 7 for Duhem's discussion of this point, in relation to the Greeks.
†Osiander's preface is discussed in relation to Galileo in Chapter 7.

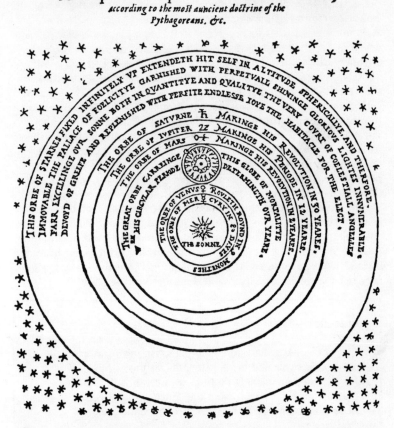

Figure 5 The Copernican System. A schematic diagram published by Thomas Digges in his 'Perfit Description of the Caelestiall Orbes', 1576. The relative sizes of the planetary orbits are not represented.

Ptolemaic system necessitated the introduction of additional assumptions.

For example, the retrogression of the planets was explained quite simply when the earth became a moving observation platform; in the Ptolemaic system major epicycles were necessary to account for this phenomenon. Additionally, observation showed that the planets Mercury and Venus were always nearer to the sun; Venus sometimes sets just after the sun, as the evening star, sometimes rises just before the sun, as the morning star. This phenomenon is referred to as limited elongation. In the Ptolemaic system it had to be explained by judicious relation of the devices governing the orbits of the sun and these planets; in the Copernican system it followed naturally from the fact that these planets orbited the sun in orbits far smaller than the Earth's, they could never therefore appear to be far away from it.

The attraction of the Copernican system as a piece of technical astronomy was that it incorporated in its basic axioms explanations which in the Ptolemaic system had to be added on. As a tool for astronomical prediction it was perhaps more pleasing to work with.

As a physical reality the Copernican system was less attractive. In the first place the idea that the earth was in motion was inconsistent with common sense and with educated opinion which was based on the teachings of Aristotle and the Church. A second reason which ran less deep but which was far from trivial was the absence of stellar parallax. If the earth is revolving around the sun it is changing its position with respect to the fixed stars and this change should be perceptible. In fact it was not; in order to account for this, Copernicus was forced to assume that the stars were immensely distant, so distant that the wisdom of the creator was challenged. If, as was believed, the stars were there for the glory of God and the entertainment of mankind what purpose could be served by situating them so far from the earth and from the orbits of the other heavenly bodies? What purpose could be served by the immense wasted space between the sphere of Saturn and the sphere of the stars?

Questions

Can you draw a diagram which demonstrates how retro-gression is explained by the Copernican theory? It is easier for the superior planets — those whose orbits are outside that of the earth. What were the other advantages of the Copernican theory? Consult the recommended texts.

Tycho de Brahe: compromise

For some decades after its publication *De Revolutionibus* was largely unread and unnoticed, although its significance as a piece of technical astronomy was gradually assumed in professional astronomical circles. The Danish nobleman, Tycho de Brahe, a pioneering observational astronomer, was forced to reject it when he discovered that stellar parallax was unobservable even with his own superb instruments which set a new standard in accuracy of naked-eye observation. Tycho generated his own system which maintained the attractive technical features of the Copernican system while leaving the earth stationary as custom and common sense demanded. In Tycho's system the sun orbited the earth, and the deferents of the planets were sun centered. It was as if in a working model of the Copernican planetary system, the sun were to be detached from the base and the earth pinned down; the relative motions of the several bodies are conserved but the earth is no longer in motion. The device is geometrically equivalent to the Copernican system and enjoys the same economical account of retrogression and limited elongation, while owing to the immobility of the earth the absence of stellar parallax is no problem and common sense is no longer violated.

Johannes Kepler

The German astronomer Johannes Kepler assisted Tycho de Brahe for a short time prior to the latter's death and eventually produced from the wealth of data that Tycho had amassed in a lifetime of observation Kepler's three laws of planetary motion which for the first time swept aside the Platonic dogma of uniform circular motion. For Kepler was able to show that the oval orbits of the planets around the sun, which Copernicus had compounded out of circles, corresponded in fact to simple ellipses with the sun at one focus. In his second law Kepler showed that the speed of a planet around this elliptical path varied such that an imaginary line drawn from the planet to the sun swept out equal areas in equal times. Kepler's third law related the average distances of the planets from the sun to the average time for the planet to complete an orbit. We note that Kepler assumed, and built on, a heliocentric model of the universe, but we do not dwell on Kepler, significant as his contribution was to the eventual Newtonian synthesis. For the focus of this book is Kepler's contemporary, Galileo, who although not unaware of Kepler's ellipses, acted for all the world as if Kepler's laws had not been revealed.

Questions

To what use would the Copernican theory be put as a piece of technical astronomy?

Can you sketch the Tychonic system?

Chapter Four
Galileo Galilei

Galileo Galilei was born in 1564 at Pisa, to a merchant family with a commitment to music, and some evidence of a patrician ancestry. It was intended that Galileo should become a doctor and he was enrolled as a student of medicine at the University of Pisa in 1581. Only at the age of nineteen was Galileo acquainted with the study of mathematics, and such was his enthusiasm that his studies in medicine were eventually abandoned in favor of mathematics and physics. After four years of private study in Florence, during which time his eventual brilliance was foreshadowed by, among other achievements, his discovery of the isochronism of small oscillations of a pendulum. Galileo returned to the University of Pisa as Professor of Mathematics.

During three years at Pisa, Galileo fulfilled an obligation to lecture on the Ptolemaic system of astronomy. The date of his commitment to the Copernican system is unknown, but there is some evidence that his private opinions at this time contrasted with the account of the Ptolemaic system which he offered in his lectures. His studies on the problem of motion during the period at Pisa, which were documented although unpublished, indicate a rejection of the Aristotelian concept of motion and a comprehension of the impetus theory of motion which is associated with the university of Paris in the fourteenth and fiteenth centuries. To Aristotle, motion other than to or from the center of the universe was an unnatural state which required the action of a force for its continuation. The impetus theory went someway towards the Newtonian conception of inertia; whereas the real motion of the earth was an absurdity to Aristotelian physics it offered no problems to a Newtonian analysis. The critical consideration by Galileo of Parisian physics is indicative, if only in its negative aspect of the rejection of Aristotle, of a comprehension within which the real motion of the earth was plausible.

It was not until 1597, five years after Galileo had moved to a chair of mathematics at the University of Padua, that his personal commitment to Copernicanism was clearly documented. In a letter to Kepler, written to acknowledge receipt of Kepler's book *Mysterium Cosmographicum* Galileo declared himself to be in agreement with Kepler about the truth of the Copernican theory. His first public commitment came seven years later when in a series of lectures on a new star Galileo argued that its existence was a strong argument in favor of the Copernican theory. In fact the existence of new stars was in opposition to Aristotle's cosmology, in which change was confined to the sub-lunary regions; as such it was indirect evidence against the Ptolemaic system but its relation to the truth or falsity of the Copernican system was no more than this.

Question

How did the Aristotelian theory of motion account for the motion of a projectile? How did it fit in with his cosmology? Use the recommended texts to find out.

Telescopic observations

In 1609 Galileo built a telescope. Popular history often accredits Galileo with its invention but its existence is recorded over a period of several years prior to 1609, in Italy, in Germany and in Holland. Galileo wrote that he constructed his telescope after hearing the instrument described. The use that he made of it was in the first instance a worldly one. He presented it to the Venetian Senate and emphasized its great military significance; as a consequence his salary was increased and his professorship confirmed for life. But the most important use that Galileo made of his telescope was to turn it to the heavens. He may not have even been the first person to do this, but the use that he made of the information that resulted makes Galileo's relationship to the telescope a unique one.

By March 1610 Galileo had published his findings in an important little book, the *Starry Messenger.* There were three important discoveries described in the *Starry Messenger.* Galileo announced that the moon was not a perfect spherical body but that it had mountains, valleys and craters, like the earth. There turned out to be stars in the heavens which were not discernible with the naked eye, in particular the Milky Way was composed of clusters of small stars. Additionally, and to Galileo most importantly, the planet Jupiter was seen to have four moons. Galileo points out that this motion of the four moons, composite with respect to the sun, will '. . . put at rest the scruples of those who can tolerate the revolution of the planets about the sun in the Copernican system, but are so disturbed by the revolution of the single moon around the earth while both of them describe an annual orbit around the sun, that they consider this theory of the universe to be impossible'.

Question

How would you answer the objection, raised in Galileo's time, that the discoveries made by the telescope were not really there, that they were produced by the distortions of the telescope?

17

ә—2

This is the only reference to the Copernican system that Galileo makes in the *Starry Messenger*. It is neither a clearly expressed commitment to the system, nor is it a conclusive or even modestly compelling argument in its favor. The role of evidence such as this in the choice between the earth-centered and sun-centered astronomies will be summarized and discussed in the next chapter.

The move to Florence

A copy of the *Starry Messenger* was sent to Cosimo de Medici, Grand Duke of Tuscany, along with the gift of a telescope and the news that Galileo had named the four moons of Jupiter the 'Medicean Planets' after the family of the Grand Duke. This personal tribute by Galileo was clearly inspired by personal motives, for Galileo's position at Padua was still financially inadequate and his teaching duties had become increasingly burdensome to him. The possibility of a post at Florence under the patronage of the Medicis was made additionally desirable by the potentially hostile scientific environment of which Galileo was becoming increasingly aware.

When Galileo was offered the post of 'Chief Mathematician of the University of Pisa and Philosopher of the Grand Duke', a post which carried a handsome salary and no teaching obligations, he immediately accepted. The condition of the post was that he should reside not at Pisa but at Florence, where he was to move in September 1610. Before he left Padua for Florence he continued his astronomical observations, recording the existence of sunspots and the strange appearance of Saturn which appeared to Galileo to have twin satellites. The first observation was to assume greater significance in 1612 when it engaged Galileo in a dispute over the priority for this discovery with the Jesuit astronomer Father Scheiner, which resulted in the publication in 1613 of *Letters on Sunspots.* Of this, more anon.

Soon after the move to Florence, Galileo was able to record his vitally important discovery that the planet Venus exhibited phases like the moon.

Animosity

The impact of Galileo's discoveries on the intellectual community of seventeenth century Europe was two-fold. In the first place there was, as might be expected, admiration for the scientist who had brought these novelties to the attention of the world. But there was also criticism from those who disputed Galileo's priority in the discovery of the telescope and in the observations that he made with it. Such disputes over priority are not uncommon in science but their frequency and intensity over this period is perhaps a pointer to the nature of the threat to astronomical theory that the dicoveries posed. For the Aristotelian/Ptolemaic world-view such observations were, if not always directly contrary to theory, at least broadly threatening.

They were anomalies which significantly threatened the authority of contemporary scholarship and the authority of the Church. The attempt to resolve these anomalies extended beyond the strictly scientific to the evidently personal. We will now discuss these anomalies, and the response that they engendered, in some detail.

Questions

At this stage, what do you think it would have taken to *prove* the Copernican theory? Would your answer have constituted conclusive proof?

Why were the phases of Venus an important discovery?

Chapter Five
The Relation of Observational Evidence to the Choice Between the Astronomical Theories

We shall discuss eight observational items which were relevant in 1612 to a choice between the competing astronomical theories of Ptolemy, Copernicus and Tycho de Brahe. The last was to assume an increasing importance over the subsequent years.

Six of these eight discoveries are attributable to Galileo, while the first two in the list had been discovered, without telescopes, by Tycho de Brahe and others in the preceding century. The list is as follows:
1. The existence of new stars, or novae.
2. The super-lunary nature of comets.
3. The existence of mountains on the moon.
4. The existence of stars which were invisible to the naked eye.
5. The moons of Jupiter — the 'Medicean Planets'.
6. The strange appearance of Saturn.
7. The existence of sun-spots.
8. The phases of Venus.

In order to understand the significance of these items it is essential to remember that the cosmology represented by the Ptolemaic system was in fact a fusion of Aristotelian physics with Ptolemaic sky-geometry. Ptolemy's universe was carried on wheels which turned within the crystal spheres of Aristotle and the whole edifice was in accord with the fundamental distinction made by Aristotle between the sub-lunary and super-lunary regions. The former contained man and was in a continuous state of change and decay, while the latter was the uncorruptible region of God's creation, a region of unchanging perfection where perfection meant sphericity.

Super-lunary change

If the super-lunary region was perfect and unchanging then no new stars should appear in the sky, and no comets should pass through this region. The moon should be a perfect sphere while the sun, too, should be spherical and unchanging. Compare now these predictions of the established cosmology with items (1), (2), (3) and (7) in the list above. It was not that these observations provided any evidence for the Copernican theory, or even against the Ptolemaic when considered as a piece of sky-geometry, but the stability of the Ptolemaic/Aristotelian cosmology taken as a whole was severely threatened.

Also, the interpretation of the sky as God's handiwork created for man's benefit was shaken by the apparent existence of heavenly bodies, stars and moons, which were invisible to man's God-given power of sight; again no fundamental contradiction of the Ptolemaic system is implied but the Christian interpretation of the Greek cosmology is questioned.

The weight of evidence such as this is therefore largely negative and even then, only indirect. A positive argument for the Copernican system was seen by Galileo in the moons of Jupiter; here were heavenly bodies which did not orbit the earth but orbited a planet much in the same way that the earth's moon was supposed in the Copernican system to orbit the earth while the earth orbited the sun. Galileo supposed that the seeming possibility of this was an argument to be leveled against the Copernican theory, and the existence of the Medicean Planets was visible proof of its weakness. The appearance of Saturn, which Galileo originally interpreted as revealing two satellites to the planet, supports the same form of argument. But the argument which Galileo opposed was an inherently weak one entirely reconcilable within a Ptolemaic framework, for any secondary satellite of the sun can be accounted for by treating the primary satellite's orbit as a deferent and carrying the secondary satellite on an additional epicycle, or set of epicycles.

The phases of Venus

If there are no clear criteria for choice in the above findings it is only necessary to look at the final piece of evidence for implications of a different nature. The planets are visible to the naked eye as mere points of light without shape. One problem which the Copernican system posed was that it predicted that Venus should be at some times sixteen times nearer the earth than at others. As Osiander had pointed out in his preface to *De Revolutionibus* 'this was . . . contradicted by the experience of all ages'. When Galileo turned his telescope to Venus he was able to discern that the planets appearance varied between a long thin crescent and a small disc — the barely variant brightness of Venus was accounted for by the exhibition of phases, a 'new' phase coinciding with a position close to the earth and a 'full' phases with a distant position.

As will be evident from *Figure 6*, the Ptolemaic system cannot account for the phases of Venus; since the planets limited elongation demands that the center of the planet's epicycle be on or near the line from the sun to earth the planet can never appear in full illumination and would always appear in 'new' or crescent form.

In the Copernican system, on the other hand, a clear explanation of the observations is exhibited. It will be recollected that the

Figure 6   Phases of Venus

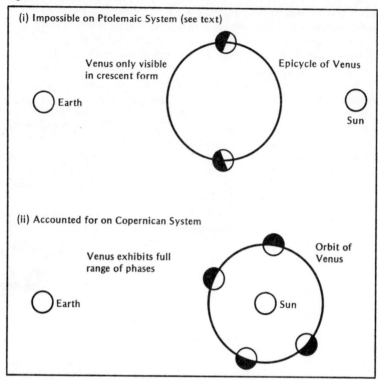

(i) Impossible on Ptolemaic System (see text)

Venus only visible in crescent form

Epicycle of Venus

Earth

Sun

(ii) Accounted for on Copernican System

Venus exhibits full range of phases

Orbit of Venus

Earth

Sun

system proposed by Tycho de Brahe preserved the geometrical harmonies of the Copernican system, it was indeed a geometrically equivalent transform to a stationary-earth frame of reference. It is evident therefore that the phases of Venus are equally well explained on the Tychonic system.

We shall now consider how Galileo's discoveries were received, and will not be surprised to find that the more open-minded astronomers, particularly in this case those of the Catholic Church, recognized that the evidence of the phases of Venus weighed heavily against the Ptolemaic system. We might not be surprised to find that the alternative that they adopted was the Tychonic compromise, which offered a full account of observation and did nor demand the apparent absurdity of a moving earth.

However, even the acceptance of this weighty piece of evidence was dependent on the belief that observations made through the telescope did correspond in status to observations made with the unaided senses. Was there a one-to-one correspondence between 'seeing' through a telescope, and just 'seeing'? How were astronomers to know that the instrument did not distort so-called reality?

Chapter Six
The Reception of Galileo's Discoveries

In assessing the reaction of Galileo's adversaries it is necessary to remember that his discoveries threatened a way of looking at the world which to its adherents was an adequate, even necessary, explanation of the human situation. Aristotelian natural philosophy and Christian theology formed a union with a solid and satisfying core of belief. Presenting an Aristotelian with evidence against his beliefs compares with confronting a modern physicist with evidence against, let us say, the theory of relativity. A research program is threatened which has already achieved a great deal of success. The natural reaction is to defend that research program; independent of any personal emotional commitment the scientist is entitled to examine critically the anomaly that confronts him and see whether it is an anomaly at all, whether it meets adequate criteria of credibility and, if so, whether it can then be explained adequately within his own set of beliefs.

Faith in the telescope

On reception of the *Starry Messenger* a number of Aristotelians reacted to the threat against their theories by refusing to admit the telescopic evidence at all. Prominent among these was Cremonini, a former friend and colleague of Galileo at Padua. In attempting to meet the criticism of men such as this Galileo was hindered by the crude nature of his instruments. His attempts to convince his opponents of the validity of telescopic observation were confounded by the simple fact that the instrument would often present the observer with a double image of a fixed star. His critics were entitled to claim that the nature of the instrument was to deceive.

Indeed it was difficult to demonstrate otherwise for the fundamental importance of the telescope was that it revealed phenomena in the heavens which were invisible to the naked eye. There was no way in which the existence of such phenomena could be independently tested.

Question

It was obviously 'unscientific' to refuse to look down the telescope, but was it 'unscientific' to argue that the observations were just distortions?

It was on the basis of faith that many Aristotelians continued to dismiss the telescopic observations and equally on the basis of faith that Galileo, and subsequently the top ranking Jesuit astronomers of the Collegio Romano, accepted the observations as fact. These latter were convinced by Galileo during a trip to Rome which he made in 1611. He was received as an honoured guest by Pope Paul V.

Galileo's faith in the telescope was based upon thousands of observations of terrestrial objects which could be subsequently verified by direct sense experience. He therefore felt justified in extending his belief to objects in the heavens for which direct verification was impossible. Given the Aristotelian assumption that the heavens were qualitatively different from the earth, such an extension could barely be justified. The development of physics since the seventeenth century leaves us in no doubt that Galileo's faith was well founded. But the basis for this faith seems, at least, to defy logical analysis.

An ad-hoc proposition

A similar question is posed by a subtle answer given by the Jesuit astronomer Clavius to the problems associated with Galileo's observations of mountains on the moon. Such observations clearly contradicted the Aristotelian orthodoxy which believed all heavenly bodies to be perfectly spherical. Clavius postulated that the craggy surface of the moon was covered by a crystalline shell, perfectly transparent but with a smooth and spherical outer surface. In a sense this argument was invincible, for the material that it postulates has no discernible properties save its sphericity. Galileo rejected it as absurd, as being 'neither demonstrated nor demonstrable'. As such it was an 'ad-hoc' proposition brought in to save the degenerating Aristotelian research program. We would have to search hard, and perhaps fruitlessly, to find methodological principles to determine what is, and what is not, an acceptable degree of scientific 'ad-hocery'. Galileo's frequent ability to recognize the fine line to be drawn between the acceptable and the unacceptable, in observation and proposition, suggests that the methods that are inevitably used have as much in common with craft skills as with formal reasoning and logical analysis.

Adversaries in the Church

The resistance to Galileo was increasingly to take a personalized form. Galileo had first observed sunspots in 1610, but had neither published nor written on the subject. Late in 1611 the Jesuit Father Scheiner made public a set of three letters on sunspots which were subsequently countered by a series of letters from Galileo — the whole set was eventually published by the Linnean Academy. Galileo made a claim for priority which he was unable to substantiate formally and his letters constituted an unpleasant attack on Scheiner. The occasion perhaps marks the first strain of a tension between Galileo and the Jesuit order which was to grow and eventually be transformed into a formal confrontation between science and the Church.

In these early days it proved to be the Dominican order rather than the Jesuit order of the Catholic Church which sharpened the conflict between the teachings of Galileo and the Church.

Friends had informed Galileo of hostile private meetings as early as 1611. In 1612 a Dominican priest, Lorini, was reported to have attacked Galileo and his views, albeit not in public and probably at the Florentine court. Galileo appears to have demanded an explanation, for a subsequent letter from Lorini to Galileo is recorded in which Lorini, in writing '. . . Ipernicus — or whatever his name is . . .' betrays an ignorance of the matter while indeed claiming no more than a conversational interest in it. Yet less than two years later the same priest was to press for a formal clerical opinion on Galileo's views on theology.

Questions at court

It is clear that Galileo's views were increasingly becoming the topic of conversation and speculation among lay members of society as well as among astronomers, philosophers and theologians. The involvement of Galileo's pupil Castelli in one such conversation initiated a sequence of events which led to Galileo's *Letter to the Grand Duchess Christina* in which Galileo made explicit his views on the relation between science and theology. Castelli was a dinner guest of Christina, mother of the Grand Duke Cosimo, on an occasion late in 1613 when the conversation turned to telescopes and astronomical observations. The consensus of the party appears to have been in favor of Galileo's telescope discoveries, but it was pointed out by a philosopher in the company that the motion of the earth was an impossibility, in particular because the Holy Scripture was obviously contrary to this view. The conversation continued after dinner and Castelli was pressed by Christina to defend the motion of the earth against the evidence of the Scriptures. Castelli was able to write to Galileo that he '. . . played the theologian with such assurance and dignity, that it would have done you good to hear me'.

Opinions in writing

Perhaps Galileo recognized at this juncture the tremendous implications which his teaching held; perhaps he had always been aware but was prompted by this event to face formally the theological problems. This he did by composing a long letter to Castelli in which he set out his opinion on the proper relations of science and theology, and included a Copernican interpretation of a particular biblical passage which seemed to imply a moving sun. This was the biblical account of the 'miracle of Joshua' of which more will presently be written.

During 1614 the attitude of the Church began to polarize. On the one hand a number of Jesuit astronomers and theologians were beginning to acknowledge publicly parts, if not the whole, of Galileo's views. There was even talk of a volume of theology which would reconcile biblical texts with Copernican teachings, and a Carmelite priest, Foscarini, was writing, unknown to Galileo, a book in support of his views which was to play a significant role in the course of events. On the other hand, a Dominican, Caccini, preached from a pulpit in Florence an attack against the new astronomy which is reported to have taken as its text 'Ye men of Galilee, why stand ye gazing up into heaven'. The play on Galileo's name was delivered in a spirit far from humorous.

Lorini, author of a previous described attack on Galileo, had acquired a copy of the *Letter to Castelli* which he clearly thought to be dangerous and heretical, and zealously forwarded it to the Holy Office. On hearing of this, Galileo forwarded a certified copy of the original (for he feared misrepresentation) to Rome, with the explanation that it had been written in haste and he intended to amplify and improve it forthwith. The amplified version was to become the *Letter to the Grand Duchess Christina*.

Early in 1615 Foscarini's book was published and a copy was sent to Cardinal Robert Bellarmine, Head of the Collegio Romano, Consultor of the Holy Office, author of the modern catechism, and the most respected theologian of the age. The book was written in defense of the Copernican system, specially against the charges that it was inconsistent with the Bible. Bellarmine wrote his opinion of the book for its author, and clearly his letter was intended for Galileo also.

Bellarmine, of course, had read Galileo's *Letters* and his text to Foscarini specifically mentions Galileo by name.

In view of Bellarmine's position the letter amounted to a statement, albeit unofficial, of the Church's attitude to the Copernican theory.

Chapter Seven
Selected Readings

Bellarmine's letter to Foscarini, 4th April, 1615

SUMMARY

Bellarmine implies that Foscarini and Galileo should treat the Copernican system as a mathematical hypothesis which accounts for the motions of the heavens. To claim that the sun is fixed and the earth is in motion violates the Scriptures, and in particular the express dictate of the Church that its own theologians should themselves decide how the bible should be interpreted. If, however, there were some proof that the Copernican hypothesis were literally true then it would be necessary to admit that the Bible was not fully understood, and use great care in explaining it. The utility of the Copernican hypothesis as a computing tool is undisputable, but there is no demonstration that in fact the sun is stationary and the earth in motion. Solomon clearly writes that the sun moves. It cannot be argued that Solomon was describing only the appearance due to the actual motion of the earth because in cases of relative motion it is possible to discern which is stationary and which is not. We clearly experience that the earth is stationary.

The main text of Bellarmine's letter follows.

First I say that it appears to me that Your Reverence and Sig. Galileo did prudently to content yourselves with speaking hypothetically and not positively, as I have always believed Copernicus did. For to say that assuming the earth moves and the sun stands still saves all the appearances better than eccentrics and epicycles is to speak well. This has no danger in it, and it suffices for mathematicians. But to wish to affirm that the sun is really fixed in the center of the heavens and merely turns upon itself without travelling from east to west, and that the earth is situated in the third sphere and revolves very swiftly around the sun, is a very dangerous thing, not only by irritating all the theologians and scholastic philosophers, but also by injuring our holy faith and making the sacred Scripture false. For Your Reverence has indeed demonstrated many ways of expounding the Bible, but you have not applied them specifically, and doubtless you would have had a great deal of difficulty if you had tried to explain all the passages that you yourself have cited.

Second, I say that, as you know, the Council [of Trent]

would prohibit expounding the Bible contrary to the common agreement of the holy Fathers. And if Your Reverence would read not only all their works but the commentaries of modern writers on Genesis, Psalms, Ecclesiastes, and Joshua, you would find that all agree in expounding literally that the sun is in the heavens and travels swiftly around the earth, while the earth is far from the heavens and remains motionless in the center of the world. Now consider whether, in all prudence, the Church could support the giving to Scripture of a sense contrary to the holy Fathers and all the Greek and Latin expositors. Nor may it be replied that this is not a matter of faith, since if it is not so with regard to the subject matter, it is with regard to those who have spoken. Thus that man would be just as much a heretic who denied that Abraham had two sons and Jacob twelve, as one who denied the virgin birth of Christ, for both are declared by the Holy Ghost through the mouths of the prophets and apostles.

Third. I say that if there were a true demonstration that the sun was in the center of the universe and the earth in the third sphere, and that the sun did not go around the earth but the earth went around the sun, then it would be necessary to use careful consideration in explaining the Scriptures that seemed contrary, and we should rather have to say that we do not understand them than to say that something is false which had been proven. But I do not think there is any such demonstration, since none has been shown to me. To demonstrate that the appearances are saved by assuming the sun at the center and the earth in the heavens is not the same thing as to demonstrate that in fact the sun is in the center and the earth in the heavens. I believe that the first demonstration may exist, but I have very grave doubts about the second; and in case of doubt one may not abandon the Holy Scriptures as expounded by the holy Fathers I add that the words *The sun also riseth, and the sun goeth down, and hasteth to the place where he ariseth* were written by Solomon, who not only spoke by divine inspiration, but was a man wise above all others, and learned in the human sciences and in the knowledge of all created things, which wisdom he had from God; so it is not very likely that he would affirm something that was contrary to demonstrated truth, or truth that might be demonstrated. And if you tell me that Solomon spoke according to the appearances, and that it seems to us that the sun goes round when the earth turns, as it seems to one aboard ship that the beach moves away, I shall answer thus. Anyone who departs from the beach, though to him it appears that the beach moves away, yet knows that this is an error and corrects it, seeing clearly that the ship moves and not the beach; but as to the sun and earth, no sage has needed to correct the error, since he clearly experiences that the earth stands still and that his eye is not deceived when it judges the sun to move, just as he is likewise not deceived when it judges that the moon and the stars move. And that is enough for the present.

Questions

What is meant by 'saving the appearances'? At this stage, do you think that science should do more than save the appearances?

Now Galileo was not speaking hypothetically in his advocacy of the Copernican system. In his *Letters* was a clear commitment to the reality of the Copernican system; Bellarmine acted as if he had never seen these letters, but of course he had and his own letter can be construed as a paternalistic warning of the inadmissibility of the arguments contained therein. A warning cast in such a fashion was likely to provoke rather than restrain a man of the temperament of Galileo.

We turn now to the problems raised by this suggested hypothetical interpretation of the Copernican theory. Bellarmine expressed his belief that Copernicus himself intended his theory to be interpreted in this way. This belief is obviously based on the preface to *De Revolutionibus* which as we know was inserted anonymously by the cleric Osiander. Whether Bellarmine knew this is not clear, and perhaps it is not important. An excerpt from the text of this preface is reproduced below. It is important because it presents an early statement of an attitude to scientific knowledge which is represented in the works of some philosophers of science up to the present day. It is an attitude to scientific knowledge which makes the Galilean campaign meaningless.

Osiander's preface to *De Revolutionibus,* 1543

SUMMARY

To declare that the earth moves and the sun is at rest is to throw established knowledge into confusion. But this work does not do this. It is not possible to know the true cause of the motions of the heavenly bodies, it is only possible to invent theories which will account for the past motions and predict future ones. This is what the author of this work has done.

TO THE READER CONCERNING THE HYPOTHESES OF THIS WORK

Since the novelty of the hypotheses of this work has already been widely reported, I have no doubt that some learned men have taken serious offence because the book declares that the earth moves, and that the sun is at rest in the centre of the universe, these men undoubtedly believe that the liberal arts, established long ago upon a correct basis, should not be thrown into confusion. But if they are willing to examine the matter closely, they will find that the author of this work has done nothing blameworthy. For it is the duty of an astronomer to compose the history of the celestial motions through careful and skilful observation. Then turning to the causes of these motions or hypotheses about them, he must conceive and devise, since he cannot in any way attain to the true causes, such hypotheses as, being assumed, enable the motions to be calculated correctly from the principles of geometry, for the future as well as the past. The present author has performed both these duties excellently. For these hypotheses need not be true nor even probable; if they provide a calculus consistent with the observations, that alone is sufficient. Perhaps there is someone who is so ignorant of geometry and optics that he regards the epicycle of Venus as probable, or thinks that it is the reason why Venus sometimes precedes and sometimes follows the sun by forty degrees and even more. Is there anyone who is not aware from this assumption it necessarily follows that the diameter of the planet in perigee* should appear more than four times, and the body of the planet more than sixteen times, as great as in the apogee†, a result contradicted by the experience of every age? In this study there are other no less important absurdities, which there is no need to set forth at the moment. For it is quite clear that the causes of the apparent unequal motions are completely and simply unknown to this art. And if any causes were devised . . . by the imagination, as indeed very many are, they are not put forward to convince anyone that they are true, but merely to provide a correct basis for calculation. Now when from time to time there are offered for one and the same motion different hypotheses (as eccentricity and an epicycle for the sun's motion), the astronomer will accept above all others the one which is the easiest to grasp. The philosopher will perhaps rather seek the semblance of the truth. But neither of them will understand or state anything certain, unless it has been divinely revealed to him. Let us therefore permit

*Perigee: point on the orbit where planet is closest to the earth.
†Apogee: point on the orbit where planet is furthest from the earth.

these new hypotheses, which are no more probable; let us do so especially because the new hypotheses are admirable and also simple, and bring with them a huge treasure of very skilful observations. So far as hypotheses are concerned, let no one expect anything certain from astronomy, which cannot furnish it, lest he accept as the truth ideas conceived for another purpose, and depart from this study a greater fool than when he entered it. Farewell.

Question

Can you comment, from your readings so far, on Osiander's observation about Venus?

The philosophy expressed by Osiander is an *instrumentalist* philosophy; it treats scientific theories as *instruments* which can be used to summarize and condense a range of past sense experiences, and, as long as they work, to predict future experiences. The instrumentalist philosophy can be justified by arguing in the following way.

When we observe the world our brain obtains its information through our senses; all that we can know are these sense experiences and not the objects that cause them. It can be assumed that there is a material reality in which the power to cause these sense experiences resides. But this is just an assumption, which instrumentalists would claim that we have no right to make. The instrumentalist would dismiss such concerns with 'reality' and just view the laws and theories of science as *aides-mémoire*, as economical ways of summarizing a host of sense experiences which can be used as instruments or tools for prediction where, like tools, they are found to work.

The French theoretical physicist, Pierre Duhem made a representative statement of the instrumentalist position in *La Théorie Physique: Son Object, Sa Structure* published in 1906 and again in an English edition in 1954 under the title *The Aim and Structure of Physical Theory*. Duhem comments from the instrumentalist position on Copernicus and Galileo, and part of his argument is reproduced below.

Pierre Duhem (1954). From *The Aim and Structure of Physical Theory*. Princeton, Princeton University Press (Reprinted by permission of Princeton University Press)

SUMMARY

Duhem points out the distinction that the Greeks made between astronomy and physics. Note the restricted meaning that he gives to the Greek use of the term physics — metaphysics, a concern with the true realities, a knowledge of which Duhem would deny. The same interpretation of astronomical theory is reflected in a quotation from Saint Thomas Aquinas and shown to agree with Osiander's preface to *De Revolutionibus* and with selected passages from Copernicus' own writing. Kepler's denial of the instrumentalist belief is cast as a '. . . naive confidence in the boundless power of the physical method', which was implicit in Galileo's campaign to have the Copernican theory accepted. Bellarmine's views are consistent with Duhem's own position.

> The Greeks clearly distinguished, in the discussion of a theory about the motion of the stars, what belongs to the physicist — we should say today the metaphysician — and to the astronomer. It belonged to the physicist to decide, by reasons drawn from cosmology, what the real motions of the stars are. The astronomer, on the other hand, must not be concerned whether the motions he represented were real or fictitious, their sole object was to represent exactly the *relative* displacements of the heavenly bodies.
>
> In his beautiful research on the cosmographic systems of the Greeks, Schiaperelli has brought to light a very remarkable passage concerning the distinction between astronomy and physics. The passage is from Posidonius, was summarized or quoted by Geminus, and has been preserved for us by Simplicius. Here it is:

> > In an absolute way it does not belong to the astronomer to know what is fixed by nature and what is in motion, but among the hypotheses relative to what is stationary and to what is moving, he inquires as to which ones correspond to the heavenly phenomena. For the principles he has to refer to the physicist.

> These ideas, expressing pure Aristotelian doctrine, inspired many a passage by the astronomers of old; Scholasticism has formally adopted them. It is up to physics — that is, to cosmology — to give the reasons for the astronomical appearances by going back to the causes themselves; astronomy deals only with the observation of phenomena and with conclusions that geometry can deduce from them. Saint Thomas, in commenting on Aristotle's *Physics*, said:

> > Astronomy has some conclusions in common with physics. But as it is not purely physics, it demonstrates them by other means. Thus the physicist demonstrates that the earth is spherical by the procedure of a physicist, for example, by

saying its parts tend equally in every direction towards the center; the astronomer, on the contrary, does this by relying on the shape of the moon in eclipses or the fact that the stars are not seen to be the same from different parts of the world.

It is by furtherance of this conception of the role of astronomy that Saint Thomas, in his commentary on Aristotle's *De Caelo*, expressed himself in the following manner on the subject of the motion of the planets:

> Astronomers have tried in diverse ways to explain this motion. But it is not necessary that the hypotheses they have imagined to be true, for it may be that the appearances the stars present might be due to some other mode of motion yet unknown by men. Aristotle, however, used such hypotheses relative to the nature of motion as if they were true.

In a passage from the *Summa Theologiae* (1, 32), Saint Thomas showed even more clearly the incapacity of physical method to grasp an explanation that is certain:

> We may give reasons for a thing in two ways. The first consists in proving a certain principle in a sufficient way, thus, in cosmology (*scientia naturalis*) we give a sufficient reason to prove that the motion of the heavens is uniform*. In the second way, we do not bring in a reason which proves the principle sufficiently, but the principle being posited in advance, we show that its consequences agree with the facts; thus, in astronomy, we posit the hypothesis of epicycles and eccentrics because, by making this hypothesis, the sensible appearances of the heavenly motions can be preserved; but that is not a sufficiently probative reason, for they might perhaps be preserved by another hypothesis.

This opinion concerning the role and nature of astronomical hypotheses agrees very easily with a good number of passages in Copernicus and his commentator Rheticus. Copernicus, notably in his *Commentariolus de hypothesibus motuum caelestium a se constitutis*, simply presents the fixity of the sun and the mobility of the earth as postulates which he asks that he be granted: *Si nobis aliquae petitiones . . . concedentur*. It is proper to add that in certain passages of his *De revolutionibus caelestibus libri sex*, he professes an opinion concerning the

---

*i.e. we make *a priori* assertions about the nature of the universe.

33

reality of his hypotheses which is less reserved than the doctrine inherited from Scholasticism and expounded in the *Commentariolus.*

This last doctrine is formally enunciated in the famous preface which Osiander wrote for Copernicus' book *De revolutionibus caelestibus libri sex.* Osiander expresses himself thus: *Neque enim necesse est eas hypotheses esse veras, imo, ne verisimiles quidem; sed sufficit hoc unum, si calculum observationibus congruentam exhibeant.** And he ends his preface with these words: *Neque quisquam, quod ad hypotheses attinet, quicquam certi ab Astronomia expectet, cum nihil tale praestare queat.†*

Such a doctrine concerning astronomical hypotheses aroused Kepler's indignation. In his oldest writing, he said:

> Never have I been able to assent to the opinion of those people who cite to you the example of some accidental demonstration in which from false premises a strict syllogism deduces some true conclusion, and who try to prove that the hypotheses admitted by Copernicus may be false and that, nevertheless, true phenomena may be deduced from them as from their proper principles . . . I do not hesitate to declare that everything that Copernicus gathered *a posteriori* and proved by observation could without any embarrassment have been demonstrated *a priori* by means of geometrical axioms, to an extent that would be a delightful spectacle to Aristotle, were he living.

This enthusiastic and somewhat naive confidence in the boundless power of the physical method is prominent among the great discoverers who inaugurated the seventeenth century. Galileo did indeed distinguish between the point of view of astronomy, whose hypotheses have no other sanction than agreement with experience, and the point of view of natural philosophy, which grasps realities. When he defended the earth's motion he claimed to be talking only as an astronomer and not to be giving hypotheses as truths, but these distinctions are in his case only loopholes created in order to avoid the censures of the church; his judges did not consider them sincere, and if they had regarded them as such, these judges would have shown

---

*Translator's note: 'Nor is it, to be sure, necessary that these hypotheses be true, or even probable; but this one thing suffices, namely, whether the calculations show agreement with the observations'.
†Translator's note: 'Nor should anyone, because he holds fast to hypotheses, expect certainty from astronomy, as it cannot be responsible for anything like that'.

very little insight. If they had thought that Galileo spoke as an astronomer and not as a natural philosopher or, in their idiom, 'physicist', if they had regarded his theories as a system suited to *represent* celestial motions and not as an affirmative doctrine about the *real nature* of astronomical phenomena, they would not have censured his ideas. We are assured of this by a letter which Galileo's principal adversary, Cardinal Bellarmine, wrote to Foscarini on April 12, 1615:

> Your Fatherhood and the honorable Galileo will act prudently by contenting yourselves to speak hypothetically, *ex suppositione*, and not absolutely, as Copernicus has always done, I believe; in fact, to say that by supposing the earth mobile and the sun stationary we give a better account of the appearances than we could with eccentrics and epicycles, is to speak very well; there is no danger in that, and it is sufficient for the mathematician.

In this passage Bellarmine maintained the distinction, familiar to the Scholastics, between the physical method and the metaphysical method, a distinction which to Galileo was no more than a subterfuge.

Question

How would you begin to argue against the instrumentalist view of science?

We must now accept the strength of Bellarmine's position: he would have allowed two interpretations of the Copernican theory, an instrumentalist one which would have denied meaning to Galileo's campaign, or an absolute one for which he would require the proof that Galileo was unable to provide. But Galileo was in a sense fighting a different battle. He took the Copernican theory to be a proven account of reality and from this basis attempted to reconcile it with Catholic teaching. In order to do so he had to make a distinction between two sorts of knowledge, scientific knowledge and theological, religious, or spiritual knowledge. The *Letter to the Grand Duchess Christina* is then the epitome of the Galileo affair, an affair which laid down the program of modern science and started it on its way at the cost of divorcing science from religion and faith from reason. Cost? Or Benefit? This can subsequently be discussed, although it is perhaps worth noting here that the growth of science-as-we-know-it is dependent on this divorce, and yet one does not have to be a devotee of institutionalized religion to appreciate that a world viewed solely as a matter in motion is not fully consistent with a world in which spiritual values such as love and beauty are fundamental human experiences.

Galileo's *Letter to the Grand Duchess Christina*

Four extracts from the *Letter* are reproduced below. Each extract is summarized separately.

1. SUMMARY

Galileo argues that the Bible is not to be taken literally, since the words of the Holy Ghost were set down by the sacred scribes in common language so that they could be understood by simple people. Particularly in physical matters, which are not the Bible's concern, confusion has been avoided by casting such casual references as are found, into simple language.

In physical matters we should use our God-given senses as the prime source of knowledge, and use this knowledge to help us understand these parts of the Bible which are not concerned with the service of God and the salvation of souls.

> The reason produced for condemning the opinion that the earth moves and the sun stands still is that in many places in the Bible one may read that the sun moves and the earth stands still. Since the Bible cannot err, it follows as a necessary consequence that anyone takes an erroneous and heretical position who maintains that the sun is inherently motionless and the earth movable.
>
> With regard to this argument, I think in the first place that it is very pious to say and prudent to affirm that the holy Bible can never speak untruth - whenever its true meaning is understood. But I believe that nobody will deny that it is often very abstruse, and may say things which are quite different from what its bare words signify. Hence in expounding the Bible if one were always to confine oneself to the unadorned grammatical meaning, one might fall into error. Not only contradictions and propositions far from true might thus be made to appear in the Bible, but even grave heresies and follies. Thus it would be necessary to assign to God feet, hands and eyes, as well as corporeal and human affections, such as anger, repentance, hatred and sometimes even the forgetting of things past and ignorance of those to come. These propositions uttered by the Holy Ghost were set down in that manner by the sacred scribes in order to accommodate them to the capacities of the common

people, who are rude and unlearned. For the sake of those who deserve to be separated from the herd, it is necessary that wise expositors should produce the true senses of such passages, together with the special reasons for which they were set down in these words. This doctrine is so widespread and so definite with all theologians that it would be superfluous to adduce evidence for it.

Hence I think that I may reasonably conclude that whenever the Bible has occasion to speak of any physical conclusion (especially those which are very abstruse and hard to understand), the rule has been observed of avoiding confusion in the minds of the common people which would render them contumacious toward the higher mysteries. Now the Bible, merely to condescend to popular capacity, has not hesitated to obscure some very important pronouncements, attributing to God himself some qualities extremely remote from (and even contrary to) His essence. Who, then, would positively declare that this principle has been set aside, and the Bible has confined itself rigorously to the bare and restricted sense of its words, when speaking but casually of the earth, of water, of the sun, or of any other created thing?Especially in view of the fact that these things in no way concern the primary purpose of the sacred writings, which is the service of God and the salvation of souls — matters infinitely beyond the comprehension of the common people.

This being granted, I think that in discussions of physical problems we ought to begin not from the authority of scriptural passages, but from sense-experiences and necessary demonstrations; for the Holy Bible and the phenomena of nature proceed alike from the divine Word, the former as the dictate of the Holy Ghost and the latter as the observant executrix of God's commands. It is necessary for the Bible, in order to be accommodated to the understanding of every man, to speak many things which appear to differ from the absolute truth so far as the bare meaning of the words is concerned. But Nature, on the other hand, is inexorable and immutable; she never transgresses the laws imposed upon her, or cares a whit whether her abstruse reasons and methods of operation are understandable to men. For that reason it appears that nothing physical which sense-experience sets before our eyes, or which necessary demonstrations prove to us, ought to be called in question (much less condemned) upon the testimony of biblical passages which may have some different meaning beneath their words. For the Bible is not chained in every expression to conditions as strict as those which govern all physical effects; nor is God any less excellently revealed in

Nature's actions than in sacred statements of the Bible. Perhaps this is what Tertullian meant by these words: 'We conclude that God is known first through Nature, and then again, more particularly, by doctrine; by Nature in His words, and by doctrine in His revealed word.'

From this I do not mean to infer that we need not have an extraordinary esteem for the passages of holy Scripture. On the contrary, having arrived at any certainties in physics, we ought to utilize these as the most appropriate aids in the true exposition of the Bible and in the investigation of those meanings which are necessarily contained therein, for these must be concordant with demonstrated truths. I should judge that the authority of the Bible was designed to persuade men of those articles and propositions which, surpassing all human reasoning, could not be made credible by science, or by any other means than through the very mouth of the Holy Spirit.

Yet even in those propositions which are not matters of faith, this authority ought to be preferred over that of all human writings which are supported only by bare assertions or probable arguments, and not set forth in a demonstrative way. This I hold to be necessary and proper to the same extent that divine wisdom surpasses all human judgement and conjecture.

But I do not feel obliged to believe that that same God who has endowed us with senses, reason, and intellect has intended to forgo their use and by some other means to give us a knowledge which we can attain by them. He would not require us to deny sense and reason in physical matters which are set before our eyes and minds by direct experience or necessary demonstrations. This must be especially true in those sciences of which but the faintest trace (and that consisting of conclusions) is to be found in the Bible. Of astronomy, for instance, so little is found that none of the planets except Venus are so much as mentioned, and this only once or twice under the name of 'Lucifer'. If the sacred scribes had any intention of teaching people certain arrangements and motions of the heavenly bodies, or had they wished us to derive such knowledge from the Bible, then in my opinion they would not have spoken of these matters so sparingly in comparison with the infinite number of admirable conclusions which are demonstrated in that science. Far from pretending to teach us the constitution and motions of the heavens and the stars, with their shapes, magnitudes, and distances, the authors of the Bible intentionally forebore to speak of these things, though all were quite well known to them. Such is the opinion of the holiest and most learned Fathers, and in St. Augustine we find the following words:

It is likewise commonly asked what we may believe about the form and shape of the heavens according to the Scriptures, for many contend much about these matters. But with superior prudence our authors have foreborne to speak of this, as in no way furthering the student with respect to a blessed life — and, more important still, as taking up much of that time which should be spent in holy exercises. What is it to me whether heaven, like a sphere, surrounds the earth on all sides as a mass balanced in the center of the universe, or whether like a dish it merely covers and overcasts the earth? Belief in Scripture is urged rather for the reason we have often mentioned; that is, in order that no one, through ignorance of divine passages, finding anything in our Bibles or hearing anything cited from them of such a nature as may seem to oppose manifest conclusions, should be induced to suspect their truth when they teach, relate, and deliver more profitable matters. Hence let it be said briefly, touching the form of heaven, that our authors knew the truth but the Holy Spirit did not desire that men should learn things that are useful to no one for salvation.

The same disregard of these sacred authors towards belief about the phenomena of the celestial bodies is repeated to us by St. Augustine in his next chapter. On the question whether we are to believe that the heaven moves or stands still, he writes thus:

Some of the brethren raise a question concerning the motion of heaven, whether it is fixed or moved. If it is moved, they say, how is it a firmament? If it stands still, how do these stars which are held fixed in it go round from east to west, the more northerly performing shorter circuits near the pole, so that heaven (if there is another pole unknown to us) may seem to revolve upon some axis, or (if there is no other pole) may be thought to move as a discus? To these men I reply that it would require many subtle and profound reasonings to find out which of these things is actually so; but to undertake this and discuss it is consistent neither with my leisure nor with the duty of those whom I desire to instruct in essential matters more directly conducing to their salvation and to the benefit of the holy Church.

From these things it follows as a necessary consequence that, since the Holy Ghost did not intend to teach us whether heaven moves or stands still, whether its shape is spherical or like a discus or extended in a plane, nor whether the earth is located at its center or off to one side, then so much the less was it intended to settle for us any other conclusion of the same kind. And the motion or rest of the earth and the sun is so closely linked with the things just named, that without a determination of the one, neither side can be taken in the other matters. Now if the Holy Spirit has purposely neglected to teach us propositions of this sort as irrelevant to the highest goal (that is, to our salvation), how can anyone affirm that it is obligatory to take sides on them, and that one belief is required by faith, while the other side is erroneous? Can an opinion be heretical and yet have no concern with the salvation of souls? Can the Holy Ghost be asserted not to have intended teaching us something that does concern our salvation? I would say here something that was heard from an ecclesiastic of the most eminent degree: 'That the intention of the Holy Ghost is to teach us how one goes to heaven, not how heaven goes'.

Question

What are your own reactions to these arguments? Think about your own views on the Bible and then assess the arguments from the point of view of the Catholic Church in the seventeenth century.

---

2. SUMMARY

Galileo now turns to the specific problem of the relation of theology to the other sciences. He begins by attacking those who would interpret the Bible lightly and superficially and use it as authority on every question of physics. Some theologians are men of 'profound learning and devout behavior' and these should not be included in the above, and yet even these men consider that they need not counter physical arguments which contradict their understanding of the Bible. They say that theology is queen of the sciences* and need not bend to accommodate any other science. But in what sense is theology queen? Not in the sense that it includes all other science, but in the sense that its subject matter is more dignified.

---

*Note the use of the word *science* as *knowledge,* from the Greek word for knowledge.

This being the case, theology should not have the power to rule over the subordinate sciences when they are concerned with physical propositions and not with supernatural things which are matters of faith.

It is obvious that (such) authors, not having penetrated the true senses of Scripture, would impose upon others an obligation to subscribe to conclusions that are repugnant to manifest reason and sense, if they had any authority to do so, God forbid that this sort of abuse should gain countenance and authority, for then in a short time it would be necessary to proscribe all the contemplative sciences. People who are unable to understand perfectly both the Bible and the sciences far outnumber those who do understand. The former, glancing superficially through the Bible, would arrogate to themselves the authority to decree upon every question of physics on the strength of some word which they have misunderstood, and which was employed by the sacred authors for some different purpose. And the smaller number of understanding men could not dam up the furious torrent of such people, who would gain the majority of followers simply because it is much more pleasant to gain a reputation for wisdom without effort or study than to consume oneself tirelessly in the most laborious disciplines. Let us therefore render thanks to Almighty God, who in His Beneficence protects us from this danger by depriving such persons of all authority, reposing the power of consultation, decision, and decree on such important matters in the high wisdom and benevolence of most prudent Fathers, and in the supreme authority of those who cannot fail to order matters properly under the guidance of the Holy Ghost. Hence we need not concern ourselves with the shallowness of those men whom grave and holy authors rightly reproach, and of whom in particular St. Jerome said, in reference to the Bible:

This is ventured upon, lacerated, and taught by the garrulous old woman, the doting old man, and the prattling sophist before they have learned it. Others, led on by pride, weigh heavy words and philosophize amongst women concerning holy Scripture. Others — oh, shame! — learn from women what they teach to men, and (as if that were not enough) glibly expound to others that which they themselves do not understand. I forbear to speak of those of my own profession who, attaining a knowledge of the holy Scriptures after mundane learning, tickle the ears of the people with affected and studied expressions, and declare that everything they say is to be taken as the law of God. Not bothering to learn what the prophets and the apostles

have maintained, they wrest incongruous testimonies into
their own senses — as if distorting passages and twisting the
Bible to their individual and contradictory whims were the
genuine way of teaching, and not a corrupt one.

I do not wish to place in the number of such lay writers some
theologians whom I consider men of profound learning and
devout behavior, and who are therefore held by me in great
esteem and veneration. Yet I cannot deny that I feel some dis-
comfort which I should like to have removed, when I hear them
pretend to the power of constraining others by scriptural authority to follow in a physical dispute that opinion which they
think best agrees with the Bible, and then believe themselves
not bound to answer the opposing reasons and experiences. In
explanation and support of this opinion they say that since
theology is queen of all the sciences, she need not bend in any
way to accommodate herself to the teachings of less worthy
sciences which are subordinate to her; these others must rather
be referred to her as their supreme empress, changing and
altering their conclusions according to her statutes and decrees.
They add further that if in the inferior sciences any conclusion
should be taken as certain in virtue of demonstrations or
experiences, while in the Bible another conclusion is found
repugnant to this, then the professors of that science should
themselves undertake to undo their proofs and discover the
fallacies in their own experiences, without bothering the
theologians and exegetes. For, they say, it does not become the
dignity of theology to stoop to the investigation of fallacies in
the subordinate sciences; it is sufficient merely for her to determine the truth of a given conclusion with absolute authority,
secure in her inability to err.

Now the physical conclusions in which they say we ought to
be satisfied by Scripture, without glossing or expounding it in
senses different from the literal, are those concerning which the
Bible always speaks in the same manner and which the holy
Fathers all receive and expound in the same way. But with
regard to these judgements I have had occasion to consider
several things, and I shall set them forth in order that I may
be corrected by those who understand more than I do in these
matters — for to their decisions I submit at all times.

First, I question whether there is not some equivocation in
failing to specify the virtues which entitle sacred theology to
the title of 'queen'. It might deserve that name by reason of
including everything that is learned from all the other sciences
and establishing everything by better methods and with pro-
founder learning. It is thus, for example, that the rules for

measuring fields and keeping accounts are much more excellently contained in arithmetic and in the geometry of Euclid than in the practices of surveyors and accountants. Or theology might be queen because of being occupied with a subject which excels in dignity all the subjects which compose the other sciences, and because her teachings are divulged in more sublime ways.

That the title and authority of queen belongs to theology in the first sense, I think will not be affirmed by theologians who have any skill in the other sciences. None of these, I think, will say that geometry, astronomy, music, and medicine are much more excellently contained in the Bible than they are in the books of Archimedes, Ptolemy, Boethius, and Galen. Hence it seems likely that regal pre-eminence is given to theology in the second sense; that is, by reason of its subject and the miraculous communication of divine relation of conclusions which could not be conceived by men in any other way, concerning chiefly the attainment of eternal blessedness.

Let us grant then that theology is conversant with the loftiest divine contemplation, and occupies the regal throne among sciences by dignity. But acquiring the highest authority in this way, if she does not descend to the lower and humbler speculations of the subordinate sciences and has no regard for them because they are not concerned with blessedness, then her professors should not arrogate to themselves the authority to decide on controversies in professions which they have neither studied nor practiced. Why, this would be as if an absolute despot, being neither a physician nor an architect but knowing himself free to command, should undertake to administer medicines and erect buildings according to his whim — at grave peril of his poor patient's lives, and the speedy collapse of his edifices.

Again, to command that the very professors of astronomy themselves see to the refutation of their own observations and proofs as mere fallacies and sophisms is to enjoin something that lies beyond any possibility of accomplishment. For this would amount to commanding that they must not see what they see and must not understand what they know, and that in searching they must find the opposite of what they actually encounter. Before this could be done they would have to be taught how to make one mental faculty command another, and the inferior powers the superior, so that the imagination and the will might be forced to believe the opposite of what the intellect understands. I am referring at all times to merely physical propositions, and not to supernatural things which are matters of faith.

Questions

When Galileo claims that scientists cannot 'not understand what they know' is he claiming certainty for science? Do you think science offers certainty?

3. SUMMARY

This extract follows on immediately from the one above. Galileo quotes Saint Augustine and interprets his words to mean that the Bible must be interpreted to correspond to soundly demonstrated physical truths, but the Bible must take precedence over physical propositions which are not soundly demonstrated. These must be held to be false; but before they are condemned they must be shown to be not rigorously demonstrated by those who oppose them. (Galileo imposes this construction on Aquinas, and is criticized by Koestler for shifting the burden of proof by a sleight of hand. Koestler's argument is reproduced below.) Attempts to show that the Copernican theory is false, have been made and have failed. In order to show it to be unsoundly demonstrated it would be necessary to ban the whole science of astronomy and to forbid men to look at the sky. And this would clearly oppose the many passages of scripture which teach us that the glory of God is to be read in the open book of nature.

> I entreat those wise and prudent Fathers to consider with great care the difference that exists between doctrines subject to proof and those subject to opinion. Considering the force exerted by logical deductions, they may ascertain that it is not in the power of the professors of demonstrative sciences to change their opinions at will and apply themselves first to one side and then to the other. There is a great difference between commanding a mathematician or a philosopher and influencing a lawyer or a merchant, for demonstrated conclusions about things in nature or in the heavens cannot be changed with the same facility as opinions about what is or is not lawful in a contract, bargain, or bill of exchange. This difference was well understood by the learned and holy Fathers, as proven by their having taken great pains in refuting philosophical fallacies. This may be found expressly in some of them; in particular, we find the following words of St. Augustine:
>
> It is to be held as an unquestionable truth that whatever

the sages of this world have demonstrated concerning physical matters is in no way contrary to our Bibles; hence whatever the sages teach in their books that is contrary to the holy Scriptures may be concluded without any hesitation to be quite false. And according to our ability let us make this evident, and let us keep the faith of our Lord, in whom are hidden all the treasures of wisdom, so that we neither become seduced by the verbiage of false philosophy nor frightened by the superstition of counterfeit religion.

From the above words I conceive that I may deduce this doctrine: That in the books of the sages of this world there are contained some physical truths which are soundly demonstrated, and others that are merely stated, as to the former, it is the office of wise divines to show that they do not contradict the holy Scriptures. And as to the propositions which are stated but not rigorously demonstrated, anything contrary to the Bible involved by them must be held undoubtedly false and should be proved so by every possible means.

Now if truly demonstrated physical conclusions need not be subordinated to biblical passages, but the latter must rather be shown not to interfere with the former, then before a physical proposition is condemned it must be shown to be not rigorously demonstrated — and this is to be done not by those who hold the proposition to be true, but by those who judge it to be false. This seems very reasonable and natural, for those who believe an argument to be false may much more easily find the fallacies in it than men who consider it to be true and conclusive. Indeed, in the latter case it will happen that the more the adherents of an opinion turn over their pages, examine the arguments, repeat the observations, and compare the experiences, the more they will be confirmed in that belief. And Your Highness knows what happened to the late mathematician of the University of Pisa who undertook in his old age to look into the Copernican doctrine in the hope of shaking its foundations and refuting it, since he considered it false only because he had never studied it. As it fell out, no sooner had he understood its grounds, procedures, and demonstrations than he found himself persuaded, and from an opponent he became a very staunch defender of it. I might also name the other mathematicians who, moved by my latest discoveries, have confessed it necessary to alter the previously accepted system of the world, as this is simply unable to subsist any longer.

If in order to banish the opinion in question from the world it were sufficient to stop the mouth of a single man —

as perhaps those men persuade themselves who, measuring the minds of others by their own, think it impossible that this doctrine should be able to continue to find adherents — then that would be very easily done. But things stand otherwise. To carry out such a decision it would be necessary not only to prohibit the book of Copernicus and the writings of other authors who follow the same opinion, but to ban the whole science of astronomy. Furthermore, it would be necessary to forbid men to look at the heavens, in order that they might not see Mars and Venus sometimes quite near the earth and sometimes very distant, the variation being so great that Venus is forty times and Mars sixty times as large at one time as another. And it would be necessary to prevent Venus being seen round at one time and forked at another, with very thin horns; as well as many other sensory observations which can never be reconciled with the Ptolemaic system in any way, but are very strong arguments for the Copernican. And to ban Copernicus now that his doctrine is daily reinforced by many new observations and by the learned applying themselves to the reading of his book, after this opinion has been allowed and tolerated for those many years during which it was less followed and less confirmed, would seem in my judgement to be a contravention of truth, and an attempt to hide and surpress her the more as she revealed herself the more clearly and plainly. Not to abolish and censure his whole book, but only to condemn as erroneous this particular proposition, would (if I am not mistaken) be a still greater detriment to the minds of men, since it would afford them occasion to see a proposition proved that it was heresy to believe. And to prohibit the whole science would be but to censure a hundred passages of holy Scripture which teach us that the glory and greatness of Almighty God are marvelously discerned in all his works and divinely read in the open book of heaven. For let no one believe that reading the lofty concepts written in that book leads to nothing further than the mere seeing of the splendor of the sun and the stars and their rising and setting, which is as far as the eyes of brutes and the vulgar can penetrate. Within its pages are couched mysteries so profound and concepts so sublime that the vigils, labors, and studies of hundreds upon hundreds of the most acute minds have still not pierced them, even after continual investigations for thousands of years. The eyes of an idiot perceive little by beholding the external appearance of a human body, as compared with the wonderful contrivances which a careful and practised anatomist or philosopher discovers in that same body when he seeks out the use of all those muscles, tendons, nerves, and bones; or when examining the functions of the heart and the other principal organs, he seeks the seat of the vital faculties, notes and observes the admirable structure of the sense organs, and (without ever ceasing in his amazement and delight) contemplates the receptacles of the imagination, the memory, and the understanding. Likewise, that which presents itself to mere sight is as nothing in comparison with the high marvels that the ingenuity of learned men discovers in the heavens by long and accurate observation. And that concludes what I have to say on this matter.

Question

Koestler claims that Galileo has 'shifted the burden of proof'. Before you read Koestler, can you assess his claim?

---

4. SUMMARY

Galileo addresses himself to the passage in the Bible which is known as the 'Miracle of Joshua'. Joshua seeks a lengthening of the day in order to prolong the course of a battle and calls out 'Sun, stand thou still'; as a consequence of divine intervention 'the sun stood still in the midst of the heavens' and the day was prolonged. A literal interpretation would seem to imply that the sun was moving and the earth stationary. Galileo provides a Copernican interpretation of the passage.

On a Ptolemaic interpretation the movement of the sun is through the ecliptic against the motion of the *primum mobile*. If this backward motion ceases, the length of the day is fractionally decreased. We would therefore have to assume that Joshua was addressing the appearances rather than the reality, and interpret the Bible accordingly. However on the Copernican system the literal meaning can be preserved. The sun rotates on its own axis and is the cause of all celestial motions (this argument follows Kepler: it is not a part of Copernicus' theory and nor does it appear to have been used before by Galileo). Therefore in calling 'Sun, stand thou still' Joshua is asking that the motion of the whole planetary system should cease.

The Bible says that the sun stood still in the midst of the heavens, and we must ask what is meant by this. If it were noon it would have hardly been necessary for Joshua to prolong the day and we can therefore assume that 'midst of the heavens' refers in fact to the center of the universe.

Now let us consider the extent to which it is true that the

famous passage in Joshua may be accepted without altering the literal meaning of its words, and under what conditions the day might be greatly lengthened by obedience of the sun to Joshua's command that it stay still.

If the celestial motions are taken according to the Ptolemaic system, this could never happen at all. For the movement of the sun through the ecliptic is from west to east, and hence it is opposite to the movement of the *primum mobile*, which in that system causes day and night. Therefore it is obvious that if the sun should cease its own proper motion, the day would become shorter, and not longer. The way to lengthen the day would be to speed up the sun's proper motion; and to cause the sun to remain above the horizon for some time in one place without declining towards the west, it would be necesarry to hasten this motion until it was equal to that of the *primum mobile*. This would amount to accelerating the customary speed of the sun about three hundred and sixty times. Therefore if Joshua had intended his words to be taken in their pure and proper sense, he would have ordered the sun to accelerate its own motion in such a way that the impulse from the *primum mobile* would not carry it westward. But since his words were to be heard by people who very likely knew nothing of celestial motions beyond the great general movement from east to west, he stooped to their capacity and spoke according to their understanding, as he had no intention of teaching them the arrangement of the spheres, but merely of having them perceive the greatness of the miracle. Possibly it was this consideration that first moved Dionysius the Aeropagite to say that in this miracle it was the *primum mobile* that stood still, and that when this halted, all the celestial spheres stopped as a consequence — an opinion held by St. Augustine himself, and confirmed in detail by the Bishop of Avila. And indeed Joshua did intend the whole system of celestial spheres to stand still, as may be deduced from his simultaneous command to the moon, which had nothing to do with lengthening the day. And under his command to the moon we are to understand the other planets as well, though they are passed over in silence here as elsewhere in the Bible, which was not written to teach us astronomy.

It therefore seems very clear to me that if we were to accept the Ptolemaic system it would be necessary to interpret the words in some sense different from their strict meaning. Admonished by the useful precepts of St. Augustine, I shall not affirm this to be necessarily the above sense, as someone else may think of another that is more proper and harmonious. But I wish to consider next whether this very event may not be understood more consistently with what we read in the Book of Joshua in terms of the Copernican system, adding a further observation recently pointed out by me in the body of the sun. Yet I speak always with caution and reserve, and not with such great affection for my own inventions as to prefer them above those of others, or in the belief that nothing can be brought forth that will be still more in conformity with the intention of the Bible.

Suppose, then, that in the miracle of Joshua the whole system of celestial rotations stood still, in accordance with the opinion of the authors named above. Now in order that all the arrangements should not be disturbed by stopping only a single celestial body, introducing great disorder throughout the whole of Nature, I shall next assume that the sun, though fixed in one place, nevertheless revolves upon its own axis, making a complete revolution in about a month, as I believe is conclusively proven in my *Letters on Sunspots*. With our own eyes we see this movement to be slanted towards the south in the more remote part of the sun's globe, and in the nearer part to tilt toward the north, in just the same manner as all the revolutions of the planets occur. Third, if we consider the nobility of the sun, and the fact that it is the font of light which (as I shall conclusively prove) illuminates not only the moon and the earth but all the other planets, which are inherently dark, then I believe that it will not be entirely unphilosophical to say that the sun, as the chief minister of Nature and in a certain sense the heart and soul of the universe, infuses by its own rotation not only light but also motion into other bodies which surround it. And just as if the motion of the heart should cease in an animal, all other motions of its members would also cease, so if the rotation of the sun were to stop, the rotations of all the planets would stop too. And though I could produce the testimonies of many grave authors to prove the admirable power and energy of the sun, I shall content myself with a single passage from the blessed Dionysius the Aeropagite in his book *Of the Divine Name*, who writes thus of the sun: 'His light gathers and converts to himself all things which are seen, moved, lighted, or heated; and in a word all things which are preserved by his splendor. For this reason the sun is called HELIOS, because he collects and gathers all dispersed things.' And shortly thereafter he says: 'This sun which we see remains one, and despite the variety of essences and qualities of things which fall under our senses, he bestows his light equally on them, and renews, nourishes, defends, perfects, divides, conjoins, cherishes, makes fruitful, increases, changes, fixes, produces, moves, and fashions all living creatures.

Everything in this universe partakes of one and the same sun by his will, and the causes of many things which are shared from him are equally anticipated in him. And for so much the more reason,' and so on.

The sun, then, being the font of light and the source of motion, when God willed that at Joshua's command the whole system of the world should rest and should remain for many hours in the same state, it sufficed to make the sun stand still. Upon its stopping all the other revolutions ceased, the earth, the moon, and the sun remained in the same arrangement as before, as did all the planets; nor in all that time did day decline towards night, for day was miraculously prolonged. And in this manner, by the stopping of the sun, without altering or in the least disturbing the other aspects and mutual positions of the stars, the day could be lengthened on earth — which agrees exquisitely with the literal sense of the sacred text.

But if I am not mistaken, something of which we are to take no small account is that by the aid of this Copernican system we have the literal, open, and easy sense of another statement that we read in this same miracle, that the sun stood still *in the midst of the heavens.* Grave theologians raise a question about this passage, for it seems very likely that when Joshua requested the lengthening of the day, the sun was near setting and not at the meridian. If the sun had been at the meridian, it seems improbable that it was necessary to pray for a lengthened day in order to pursue victory in battle, the miracle having occurred around the summer solstice when the days are longest, and the space of seven hours remaining before nightfall being sufficient. Thus grave divines have actually held that the sun was near setting, and indeed the words themselves seem to say so: *Sun, stand thou still, stand thou still.* For if it had been near the meridian, either it would have been needless to request a miracle, or it would have been sufficient merely to have prayed for some retardation. Cajetan is of this opinion, to which Magellan subscribes, confirming it with the remark that Joshua had already done too many things that day before commanding the sun to stand still for him to have done them in half a day. Hence they are forced to interpret the words in the *midst of the heavens* a little knottily, saying that this means no more than that the sun stood still while it was in our hemisphere; that it, above our horizon. But unless I am mistaken we may avoid this and all other knots if, in agreement with the Copernican system, we place the sun in the 'midst' — that is, in the center — of the celestial orbs and planetary rotation, as it is most necessary to do. Then take any hour of the day, either noon, or any hour as close to evening as you please, and the day would be lengthened and all the celestial revolutions stopped by the sun's standing still *in the midst of the heavens;* that is, in the center, where it resides. This sense is much better accommodated to the words, quite apart from what has already been said; for if the desired statement was that the sun was stopped at midday, the proper expression would have been that it 'stood still at noonday' or 'in the meridian circle', and not 'in the midst of the heavens'. For the true and only 'midst' of a spherical body such as the sky is its center.

Question

How in this extract does Galileo use the arguments set out in the first extract?

We can now look at the opinions of two modern writers on the subject of the *Letter to the Grand Duchess Christina*. The first is the novelist and intellectual Arthur Koestler, the second the Italian philosopher of science, Ludovico Geymonat.

Koestler writes as a critic of the divorce of faith and reason for which he, at least partly, blames Galileo. Moreover he writes as an enthusiastic biographer of Kepler in whom he sees both mystic and scientist unhappily astride the watershed of an old rationality and a new. Galileo's behavior towards Kepler — whose work and much of whose correspondence he ignored — gives Koestler a self-confessed bias in his attitude to Galileo. In the following extract he comments on the 'sophistry, evasion and plain dishonesty' which are to be found in the *Letter to the Grand Duchess Christina* along with 'superb formulations in defence of the freedom of thought'.

Arthur Koestler (1959). *The Sleepwalkers.* London, Hutchinson, pp. 436—439 (also in Penguin, 1964) (Reprinted by permission of A.D. Peters & Co Ltd)

While reading this superb manifesto of the freedom of thought, one tends to forgive Galileo his human failings. These, however, become only too apparent in the piece of special pleading which follows the passage I have quoted, and which was to have disastrous consequences.

After invoking Augustine's authority once more, Galileo draws a distinction between scientific propositions which are 'soundly demonstrated' (i.e. proven) and others which are 'merely stated'. If propositions of the first kind contradict the apparent meaning of passages in the Bible, then, according to theological practice, the meaning of these passages must be reinterpreted — as was done, for instance, with regard to the spherical shape of the earth. So far he has stated the attitude of the Church correctly; but he continues: 'And as to the propositions which are stated but not rigorously demonstrated, anything contrary to the Bible involved by them must be held undoubtedly false and should be proved so by every possible means.'

Now this was demonstrably not the attitude of the Church. 'Propositions which are stated but not rigorously demonstrated,' *such as the Copernican system itself*, were not condemned outright if they seemed to contradict Holy Scripture; they were merely relegated to the rank of 'working hypotheses' (where they rightly belong), with an implied: 'wait and see; if you bring proof, then, but only then, we shall have to reinterpret Scripture in the light of this necessity.' But Galileo did not want to bear the burden of proof; for the crux of the matter is, as will be seen, that he had no proof. Therefore, firstly, he conjured up an artificial black-or-white alternative, by pretending that a proposition must either be accepted or outright condemned. The purpose of this sleight of hand becomes evident from the next sentences.

> Now if truly demonstrated physical conclusions need not
> be subordinated to biblical passages, but the latter must
> rather be shown not to interfere with the former, then
> *before a physical proposition is condemned it must be
> shown to be not rigorously demonstrated* — and this is to be
> done not by those who hold the proposition to be true, but
> by those who judge it to be false. This seems very reasonable
> and natural, for those who believe an argument to be false
> may much more easily find the fallacies in it than men who
> consider it to be true and conclusive . . .

The burden of proof has been shifted. The crucial words are those in (my) italics. It is no longer Galileo's task to prove the Copernican system, but the theologian's task to disprove it. If they don't, their case will go by default, and Scripture must be reinterpreted.

In fact, however, there had never been any question of condemning the Copernican system as a working hypothesis. The biblical objections were only raised against the claim that it was *more* than a hypothesis, that it was rigorously proven, that it was in fact equivalent to gospel truth. The subtlety in Galileo's manoeuvre is that he does not explicitly raise this claim. He cannot do so, for he had not produced a single argument in support of it. Now we understand why he needed his black-or-white alternative as a first move: to distract attention from the true status of the Copernican system as an officially tolerated working hypothesis awaiting proof. Instead, by slipping in the ambiguous words 'physical proposition' at the beginning of the italicized passage, followed by the demand that 'it must be shown not to be rigorously demonstrated', he implied (though he did not dare to state it explicitly) that the truth of the system *was* rigorously demonstrated. It is all so subtly done that the trick is almost imperceptible to the reader and, as far as I know, has escaped the attention of students to this day. Yet it decided the strategy he was to follow in coming years.

Throughout the document Galileo completely evaded any astronomical or physical discussion of the Copernican system; he simply gave the impression that it was proven beyond doubt. If he had talked to the point, instead of around it, he would have had to admit that Copernicus' forty-odd epicycles and eccentrics were not only not proven but a physical impossibility, a geometrical device and nothing else; that the absence of an annual parallax, i.e. of any apparent shift in the position of the fixed stars, in spite of the new telescope precision, weighed heavily against Copernicus; that the phases of Venus disproved Ptolemy, but not Herakleides or Tycho; and that all he could claim for the Copernican hypothesis was that it described certain phenomena (the retrogression) more economically than Ptolemy; as against this, the above-mentioned physical objections would have carried the day.

For it must be remembered that the system which Galileo advocated was the orthodox Copernican system, designed by the Canon himself, nearly a century before Kepler threw out the epicycles and transformed the abstruse paper-construction into a workable mechanical model. Incapable of acknowledging that any of his contemporaries had a share in the progress of astronomy, Galileo blindly and indeed suicidally ignored Kepler's work to the end, persisting in the futile attempt to bludgeon the world into accepting a Ferris wheel with forty-eight epicycles as 'rigorously demonstrated' physical reality.

What was the motive behind it? For almost fifty years of his life, he had held his tongue about Copernicus, not out of fear to be burnt at the stake, but to avoid academic unpopularity. When, carried away by sudden fame, he had at last committed himself, it became at once a matter of prestige to him. He had said that Copernicus was right, and whosoever said otherwise was belittling his authority as the foremost scholar of his time. That this was the central motivation of Galileo's fight will become increasingly evident. It does not exonerate his opponents; but it is relevant to the problem whether the conflict was historically inevitable or not.

The final section of the *Letter to the Grand Duchess* is devoted to the miracle of Joshua. Galileo first explains that the sun's rotation around its axis is the cause of all planetary motion. 'And just as if the motion of the heart should cease in an animal, all other motions of its members would also cease, so if the rotation of the sun were to stop, the revolutions of all the planets would stop too.' Thus he not only assumed, with Kepler, the annual revolutions of the planets to be caused by the sun, but also their *daily* rotation round their axes — an *ad hoc* hypothesis with no more 'rigorous proof' than the analogy with the animal's heart. He then concludes that when Joshua cried: 'Sun, stand thou still,' the sun stopped rotating, and the earth in consequence stopped both its annual and daily motion. But Galileo, who came so close to discovering the law of inertia, knew better than anybody that if the earth suddenly stopped dead in its track, mountains and cities would collapse like match-boxes; and even the most ignorant monk, who knew nothing about impetus, knew what happened when the horses reared and the mail-coach came to a sudden halt, or when a ship ran against a rock. If the Bible was interpreted according to Ptolemy, the sudden standstill of the sun would have no appreciable physical effect, and the miracle remained credible as miracles go; if it was interpreted according to Galileo, Joshua would have destroyed not only the Philistines, but the whole earth. That Galileo hoped to get away with this kind of painful nonsense, showed his contempt for the intelligence of his opponents.

In the *Letter to the Grand Duchess Christina* the whole tragedy of Galileo is epitomized. Passages which are classics of didactic prose, superb formulations in defence of the freedom of thought, alternate with sophistry, evasion and plain dishonesty.

Geymonat writes as a professional philosopher whose concern is in part with the methodological revolution brought about by Galileo, and the philosophical assumptions which underlie it. He writes of 'the wonderful richness of Galileo's thought and the decisive significance of his battle in the history of modern culture', and claims an objectivity in his exposition which Koestler denies in his own. The text below concerns the Galilean distinction between scientific language and ordinary language, and the alleged primacy of observation and reason over the Bible in matters concerned not with religion but with physics.

L. Geymonat (1965). *Galileo Galilei.* New York, McGraw Hill, pp. 68—71 (Reprinted by permission of Giulio Einaudi Editore)

SUMMARY

Geymonat points out the political (and ethical?) implications of the distinction between the two languages. For the distinction between languages is developed by Galileo into a disticntion between the disciplines of theology and science.

But where is the definition of science to end? Could it not be possible that 'sense experience and reason' could yield answers to moral and ethical problems? What guarantee could Galileo offer that such a development would not ensue, that the freedom he sought for science would not open the door to a freedom destructive to the faith?

Moreover the precision that Galileo attributed to scientific language made it immune to any criticism in ordinary language whether from religion or from elsewhere. Science thus claims an absolute authority and scientific conclusions gain the status of irrefutable truths.

Galileo's solution is thus so simple (perhaps so simple-minded) that it can hardly fail to leave us perplexed. The very fact that the ecclesiastical authorities of his time would not let themselves be persuaded by it strengthens our own perplexity. However, we shall limit ourselves to those objections that can be drawn from Galileo's own subsequent development of his basic propositions. The first and most important such development consists in changing the distinction between two types of discipline: on the one hand, ethics and religion; on the other, physics.

In the first of these fields, Galileo concedes without debate that their object 'transcends all human reasoning,' by which it is clear that he means a scientific reasoning. He concedes as an obvious consequence that the truth of such disciplines 'cannot be made credible to us by any other science nor by any other means than by the mouth of the Holy Spirit'. These truths are indeed of interest to every man, he says, since they bear upon the salvation of our souls; therefore the Holy Spirit was obliged to express itself in ordinary language, which is the only one that everyone understands.

In the second field, however, Galileo maintains that man possesses, as a gift from God himself, certain faculties naturally suited for the discovery of truth with scientific rigor. Especially for that reason, he therefore raises the following question, rather a leading question: Why should God, not content with

55

the natural means he had supplied to men, add a supernatural means, and thus reveal that same truth to him through the Sacred Scriptures? Galileo had no doubt as to the answer. It appeared to him impossible that God wished to carry out so superfluous an act, or that 'that same God who endowed us with senses, with reason, and with intellect wished us to put off their use, giving us by other means the information we could acquire through them.' This would be still more absurd when one considers that physical entities can supply us with this information in a perfectly rigorous language, while the Sacred Scriptures would furnish it only with all the imprecisions of ordinary language.

But here a great difficulty arises, which even Galileo seems not to have taken fully into account: Who can deny that the reason and intellect with which we are endowed, if not the senses, may sooner or later turn out to be able to handle with scientific rigor those truths which concern the moral disciplines? It is true that Galileo says repeatedly and clearly that one needs complete freedom from the Bible only for scientific knowledge and concerning 'debates over physical problems,' but what guarantee could he offer to the Church that others who would follow the path he had opened would not require an analogous liberty for debates on moral or religious problems? What guarantee could he offer that the method of scientific research, once it emerged victorious in the field of physics, would not seek to extend itself also into ethics and religion? From their point of view, the theologians were perfectly right; they foresaw and feared a situation so dangerous to them. Nor does it matter whether the danger was immediate or remote.

While on the one hand, as Galileo's adversaries well knew, he seemed to recognize the equal right of the two languages, on the other he entertained no doubt whatever concerning the incontestable superiority of the scientific language of rigorous research over that of everyday life and of the Bible. The foundation of his thought was this: when a question has been dissected by scientific reasoning, any wish to refute the results thus arrived at by invoking statements from ordinary language loses all sense, and it makes no difference whether such statements are made by men in everyday life or by the Holy Spirit in the Bible. Confronted with truths demonstrated by science, ordinary language has absolutely nothing to say against them. A single example sufficed. In the past, some objections had been raised against the Medicean planets on the basis of Holy Writ: 'Now that everyone has seen these planets, I should like to know what new interpretations these same antagonists employ in expounding the Scriptures.'

Thus, according to Galileo's conception, scientific reasoning possesses an intrinsic value which is incontestable, and which does not have to be supported on any authority extraneous to itself. But ordinary reasoning has a limited value; on any given problem, there is no way out but to use the latter as the basis for interpreting the former. This held for the Medicean planets, and in the same way it held for Copernicanism. Narrow-minded theologians who wanted to limit science on the basis of biblical reasoning would do nothing but cast discredit upon the Bible itself.

Several years later, on the eve of his departure for Rome at the summons of the Holy Office, Galileo was to repeat once more this same worry of his (in a letter to Elia Diodati dated January 15, 1633): 'If Fromond or others have established that to say the earth moves is heresy, and if [later] demonstrations, observations, and necessary experiences show it to move, in what predicament will they have placed the Church itself? All these energetic declarations help us to understand that the atonomy of scientific thought defended by Galileo was really very broad; in fact, he required the recognition not only of the liberty of science to reject the dictates of some other form of knowledge, but also of its task of giving the unique and definite criterion of truth to all other forms of knowledge. What sense is there, he asked, in denying this autonomy when scientific reasoning is based not on mere whims, but on verifiable facts, possessing a force that no man can bend to his will? 'If the earth moves *de facto*, we cannot change nature and have it not move'.

Chapter Eight
Postscript: Galileo and the Copernican System After 1615

The *Letter to the Grand Duchess Christina* sums up the Galilean program; it has been set in the historical context of which it was a product.

The course of events between 1615 and Galileo's trial is of direct relevance to the history of science, but bears a secondary relationship to our study of the implications of the methodology of Galilean science. Superficially, at least, Galileo was defeated by the Inquisition. But it is clear to all of us that his science survived; the main thrust of science since Galileo has followed the pattern he set. The empirical method has ousted *a priori* argument and religious revelation from an ever-widening range of human experience; the notion of purpose as a component of explanation has been ousted even from biology and finds a last refuge from mechanical causality in the social sciences. The Galilean program has reigned supreme.

We can therefore discuss the implications of the Galilean program on the basis of our general knowledge or from a study of the history of modern science. Yet the demands of historical integrity must ensure that events after 1615 are discussed in this book. Therefore a brief account of the course of events is given below; the details can be obtained from the reading list provided.

The letters to Castelli and Christina had been brought to the attention of the Inquisition by Caccini but by November 1615, largely owing to the poor quality of Caccini's evidence, all charges against Galileo were dropped. It was in December 1615 that Galileo traveled to Rome for the next round of his battle. He took with him a new weapon, his account of the tides, which he believed constituted the desired proof of the earth's motion. Briefly, this argument depended on the idea that annual revolution and diurnal rotation were alternately opposed and super-imposed at any one point on the earth, the resultant effect being to make the oceans slop about. To judge it harshly: it predicted only one tide per day whereas there were two, and a moments thought shows that it contradicts Galileo's own researches into dynamics.

Galileo was not well received on this visit to Rome, and eventually matters came to a head when the Pope summoned the theologians of the Holy Office to give a formal opinion on the two following propositions:

1. The sun is the center of the world and wholly immovable of local motion.

2. The earth is not the center of the world nor immovable, but moves as a whole, also with diurnal motion.

The Qualifiers gave the verdict that the first proposition was formally heretical and the second erroneous in faith, but this verdict was not published in full untill 1633. Instead a more moderate decree was issued which did not mention heresy, instead it suspended *De Revolutionibus* pending correction and prohibited Foscarini's book, decreeing the Copernican doctrine '. . . altogether opposed to Holy Scripture'. Galileo was not mentioned in the decree which was published on 5th March, 1616. But a few weeks earlier he had been summoned to a private meeting with Bellarmine where he appears to have been formally warned of the consequences of holding his opinions. There is some doubt as to the precise wording of the warning, and it is not clear whether Galileo was just forbidden to hold and defend Copernican views or actually forbidden to discuss them. The minutes of the meeting, and a note from Bellarmine to Galileo, suggests that the weaker form of the warning was delivered; but another minute in the Vatican files implies that the freedom even to discuss the Copernican System was removed. This became critical at the time of Galileo's trial in 1633.

For a time Galileo was silent on the subject of astronomy, but in 1618 the appearance of three comets in rapid succession opened up the old arguments. It was of course Tycho de Brahe who had first demonstrated the super-lunary nature of comets, and Jesuit astronomers who by now accepted the Tychonic compromise system also accepted Tycho's views on comets. We have seen that the acceptance of compromise was alien to Galileo's nature, and in particular that he rejected absolutely the Tychonic system. Thus he was led to reject the Tychonic and Jesuit interpretation of the nature of comets in an important work *The Assayer.* It was published in 1623, and although it reverted to an earlier explanation of comets as meteorological phenomena or optical illusions it contains some typically brilliant statements of the principles of scientific reasoning. Of course it served once more to widen the gulf between Galileo and the Jesuits.

Yet in 1623 a new Pope was elected; a man who in 1616 had opposed the decree against Copernicus and whose admiration for Galileo was openly evident. Cardinal Barberini, as Urban VIII, was potentially a valuable ally to the Galilean cause and Galileo at last felt able to embark on his major treatise on the Copernican system. It was cast as a dialogue between three characters, a spokesman for Galileo, an intelligent layman and a defender of Aristotle and Ptolemy. In *The Dialogue on the Great World Systems,* as in many other polemic works in dialogue form, the character opposed to the author's views never really stood a chance. This character, Simplicio, was shown up as a fool time and time again and the dialogue ends with Simplicio putting forward an instrumentalist view of the Copernican hypothesis which was virtually identical to the position expounded by Urban. The other two characters are silenced by this argument in what is almost a display of mock reverence.

Thus it was not difficult for Galileo's enemies to convince Urban that a slight had been perpetrated against him. Moreover, the book, finished in manuscript form in 1630, was published in 1632 in a way which made it appear at least possible that Galileo had deliberately outwitted the Church's censors. Thus, early in 1633, Galileo was brought to trial by the Inquisition. The formal warning of 1616 was submitted as evidence, and the version proffered which forbade Galileo to '. . . hold, defend or teach that opinion in any way whatsoever', i.e. the version which prohibited even discussion of the Copernican theory. Galileo claimed to be unaware of the key words in that injunction, but Bellarmine was dead, and the formal record existed; its dictates were clearly countered by the *Dialogue* and the outcome of the trial was barely in question.

Contrary to popular account, Galileo was never tortured. Rather, as an old man of seventy, he was persuaded to formally recant his opinions. This he did. Contrary to popular account, he did not finally strike the ground with his foot and cry *'eppur si muove'* (and yet it does move). His book was prohibited and he was confined to '. . . formal prison during the Holy Office's pleasure'. Formal prison amounted to a series of comfortable suites in approved villas and palaces. During the next few years Galileo, by now nearly blind, returned to the study of dynamics and published *Dialogues Concerning Two New Sciences* on which his fame as a scientist was to rest. He died in 1642, the year in which Newton was born.

Chapter Nine
Notes on the Use of this Material

Introduction

The preceding text of this book consists of a historical account of the Copernican Revolution, concentrating on the part played by Galileo and including several readings either extracted from his *Letter to the Grand Duchess Christina* or of further relevance to these extracts. The intention of that text is to provide the historical material which will serve as a concrete example for an introductory-level discussion of the nature of scientific knowledge and its relation to society. These notes contain some additional guidance which might be of use in turning the case study into a teaching vehicle.

Key issues defined and structured

A breakdown of the key issues (1, 2 and 3) into sub-problems and specific questions is offered below.

1. What is the nature and meaning of scientific knowledge? Specifically: What interpretation was put on scientific knowledge by Galileo? Is this interpretation justified?

    (a) Is scientific knowledge instrumentalist, or descriptive of reality?

      (i) What are the implications of Osiander's preface?
      (ii) Does this correspond to Galileo's view?
      (iii) What additional assumptions are implicit in the Galilean view?
      (iv) Are these assumptions justified?

    (b) Is scientific knowledge certain?

      (i) Was the Copernican system 'proven'?
      (ii) What would have constituted a proof?
      (iii) What tests of acceptability exist?
      (iv) Are any of these logically watertight?
      (v) How do major changes in theory (scientific revolutions) occur?

    (c) Is the authority of science absolute?

      (i) Is it fair to assume that Galileo was rational and the Church irrational?

2. How does science relate to its external environment? Specifically: What constraints were imposed on science and on the Church by their interrelationship, and how was the tension resolved?

(a) Does scientific knowledge have social implications?

    (i) What were the social implications of the Copernican theory?
    (ii) Was the Church's reaction to the Copernican theory 'unscientific'?
    (iii) Did Galileo recognize the social implications of his science?
    (iv) How can Galileo's self-defined role as science advisor to the Church, be judged?
    (v) How does *Letter to the Grand Duchess Christina* stand up as a science advisor's document?

(b) Do external factors affect the progress of scientific knowledge?*

    (i) Did the Copernican Revolution arise out of the internal dynamic of scientific progress, or was it to some extent a result of external factors?
    (ii) Did the involvement of the Church substantially affect the outcome between Ptolemaic and Copernican astronomy?
    (iii) Was Galileo's science substantially modified by his interaction with the Church?
    (iv) In what other ways was Galileo's science influenced by factors external to science?

3. What are the implications of Galileo's definition of scientific knowledge for a complete understanding of human experience? Do different definitions of scientific knowledge carry different implications?

(a) Does a scientific approach tell us *all* that we can know about the world we live in?

    (i) Did Galileo claim this?
    (ii) Does Galileo's definition of science imply this?
    (iii) Is this implicit in eighteenth century materialist views of science summed up Laplace in the notion that the whole future of the universe could in principle be predicted if the mass, velocity and position of every particle in it were known?
    (iv) Are the implications of an instrumentalist view of science in accord with his view?

(b) Can a formal distinction be made between human experience such as love and beauty, and human experiences of mass, length and time?

*A full treatment of this question is outside the intended scope of this book.

(i) Did Galileo believe this?
(ii) What assumptions underlie this view?
(iii) Is a denial of this view necessarily anti-rational, anti-scientific?

Notes and a short bibliography on the background to the key issues

This section takes in turn each of the problem-areas defined above and suggests further reading which will provide background material. References are to the bibliography at the end of this book.

1(a) IS SCIENTIFIC KNOWLEDGE INSTRUMENTALIST, OR DESCRIPTIVE OF REALITY?

Readings:
(i) Preceding text on Osiander and Duhem (*see* Chapter 7).
(ii) Duhem (1954) especially Part 1, Chapters 1, 2, also foreword by Louis de Broglie
Duhem wishes to rid science of metaphysics, to demonstrate a view of scientific theories in which 'agreement with experiment is the sole criterion of truth for a physical theory', so that everyone can give assent to theory regardless of metaphysical assumptions. If theory is to claim to reveal ultimate reality it must be subordinated to meta-physics, to philosophical and untestable assumptions about the basis of physical experience. To establish the autonomy of physical science we have to see physical theory as just 'a set of mathematical propositions whose aim is to represent, as completely and as exactly as possible, a group of experimental laws'. Thus theories are just instruments which operate according to the principle of intellectual economy and enable us to summarise a host of physical sense-experiences.
(iii) Popper (1963) especially Chapters 1 and 3
Chapter 3 discusses three views of human knowledge and bases the discussion on the Galileo case. Galileo's view that the Copernican theory was simply a true description of the world is described as essentialist, and this is contrasted with the instrumentalist position of Osiander and Bellarmine. Popper's own position constitutes the third view. He dismisses essentialism and instrumentalism largely on the grounds of 'obscurantism', i.e. each will eventually restrict or obscure scientific progress. His own view is that scientific theories are conjectures about the world which can come into contact with the reality of that world when they are submitted to severe tests. Popper does not believe in any ultimate reality that we can know, but in a many layered reality — everyday objects, their material constituents, their molecules and atoms, and their quantized fields of forces are levels of reality which to Popper are equally conjectural

(i.e. they are descriptive categories imposed by the human mind) but are, in his terms, equally real. The crucial thing to Popper is that 'if a theory is testable, then it implies that events of a certain kind cannot happen; and so it asserts something about reality'.

(iv) Kuhn (1962, and 2nd edition 1970)

This is probably the most important book on the nature of scientific knowledge to appear in recent years. As such, its content is relevant to many of the issues discussed in this book and deserves complete and close attention.

Essentially, Kuhn believes that scientific theories cannot be logically verified (and here he is in agreement with most modern philosophers of science, including Popper) but also that scientific theories cannot be logically refuted (and here he is in disagreement with Popper, at least at the level at which we are able to present Popper in this book — see sub-problem 1(b), below). As a consequence, changes in scientific theories (or paradigms as Kuhn calls them — theories plus assumptions, tacit and explicit, about the ways in which these theories can be used) can never be logically compelling but depend on a shift in the consensus of the scientific community. In choosing a scientific theory we choose the most satisfactory set of conceptual boxes into which to force our experiences of the world. The choice between one set of boxes, one paradigm, and another is likened to the selection of one particular view of a Gestalt* diagram. But nature, unlike the Gestalt analogy, provides no neutral data (no black lines on white paper), and the alternative to viewing nature through one paradigm is viewing nature through another paradigm. As a consequence, Kuhn is able to claim that after a change in paradigm (a scientific revolution) scientists are living in a different world. The notion of 'reality', at least in its naive form as 'that which lies beneath sensible appearance', is removed once and for all as a key problem of scientific methodology. 'Truth' is no longer an issue and criteria for paradigm selection become guided by personal commitment pragmatic in a sense which is reminiscent of the instrumentalist tradition.

It is worth noting that the instrumentalist (or the positivist, a virtually synonymous term) takes the basic data on which theories are constructed as absolute, as unproblematic. The advance of the Kuhnian analysis is that observation is now recognized as theory laden and theories can no longer command universal assent. This analysis has served to fuse the sociological and philosophical analysis of science, for scientific theory is recognized as being potentially value laden (cf. Duhem's attempt to relegate metaphysics) and the

---

*Exemplified by a line drawing which the observer can interpret in more than one way.

64

cognitive content of science open to social determination and sociological analysis.

1(b) IS SCIENTIFIC KNOWLEDGE CERTAIN?

Readings:

(i) Popper (1963) especially Chapter 1

The root of this issue lies in David Hume's famous problem of induction. Hume believed that we draw generalizations from experience, but in order to be certain that these generalizations will continue to be valid in the future we have to make an unwarranted assumption about the causes of our experience. Popper sets up an alternative view of knowledge — rather than deriving our generalizations from our experiences we impose generalizations upon the world and continue to hold them only until evidence forces us to reject them. The process is one of *conjecture* and *refutation*. Only the falsity of a theory can be inferred from empirical evidence.

Thus we cannot prove a scientific theory to be true, because we have no reason to believe that it will continue to pass the tests with which it is faced, we can only prove it to be false (but see below). Scientific knowledge is therefore not certain.

(ii) Kuhn (1962)

See notes above. Kuhn recognizes that we cannot even be certain that we have refuted a theory, since the process of refutation cannot be logically compelling. The problems are illustrated by Lakatos, in Lakatos and Musgrave (1970), with a hypothetical account which in summary reads thus:

> It is discovered that the orbit of a planet deviates from that predicted by Newton's laws. Are Newton's laws refuted? No it must be that another small planet exists which is perturbing the orbit. But telescopic observations do not show this planet. Are Newton's laws refuted? No, it must be that a cloud of cosmic dust is hiding the planet. But a satellite cannot find this cloud. Are Newton's laws refuted? No, it must be that a magnetic field is disturbing the instruments — and so the story can go on, ad infinitum.

As Kuhn writes, 'No process yet disclosed by the historical study of scientific development at all resembles the methodological stereotype of falsification by direct comparison with nature'. Kuhn recognizes that scientists treat apparent counter-instances as puzzles which they assume that they will be able to solve within the current paradigm. Thus he defines *normal science* as puzzle solving. Only under certain circumstances will normal science break down and a period of crisis result in which a scientific revolution may take place. These circumstances will include a build up of outstanding puzzles, but external factors may play a significant role in the promotion of a crisis situation.

Thus the certainty of scientific knowledge is denied by Kuhn, just as it was denied by Popper. But Kuhn goes beyond the Popperian account by recognizing that even the process of refutation is not logically certain. Adherence to a scientific paradigm depends on the maintenance of an adequate puzzle-solving tradition.

It is worth noting that the position of Popper as a 'naive falsificationist is consistent with the readings above and consistent with the position of many of his advocates and commentators. In fact his writings reveal a much more sophisticated viewpoint which is in many aspects in accord with the Kuhnian position. But a discussion at that level is outside the scope of this book. Those who wish to understand more fully the Popperian position might start by reading Lakatos (op. cit.).

1(c) IS THE AUTHORITY OF SCIENCE ABSOLUTE?

The answers to this question largely follow from the readings in 1(a) and 1(b) above. If we accept the Kuhnian viewpoint we see that science comprises a consensual imposition of explanatory categories upon human experiences. We can confidently call the results of this process knowledge, but since it is capable of reflecting ideological positions, and since it is consensual rather than certain knowledge, we can confidently dispute its ability to claim authority in all situations. (See also discussion and readings under (3) below).

2(a) DOES SCIENTIFIC KNOWLEDGE HAVE SOCIAL IMPLICATIONS?

The answer to this question is self evidently yes. Science, through technology, obviously changes the social conditions of human existence. Changes in scientific theory which result in a new view of the world (to Kuhn, a new world) will often come into conflict with socio-political norms which premise the old view of the world. The Galileo affair is a case in point. The contemporary controversy over the inheritance of intelligence and its implied variation between races has obvious political implications in relation to educational programs.

Background reading for these issues is necessarily diffuse. Perusal of popular science journals (e.g. *New Scientist*) will provide plenty of evidence of the social impact of science and technology. The suggested reading will provide an introductory overview of the general themes.

Reading:
(i) Jevons (1973) especially Chapters 1, 6
A useful introduction to the problems of the social relations of science, emphasizing the connection between scientific knowledge

and largely contemporary socio-technological issues such as war, pollution and the creation of economic wealth.

2(b) DO EXTERNAL FACTORS AFFECT THE PROGRESS OF SCIENTIFIC KNOWLEDGE?

The issue at stake here is whether the progress of science is determined by its own internal logic, or by external factors. The issue is doubtless one of degree, and the Kuhnian analysis provides a framework within which to study the balance of internal and external factors. Given a situation of normal science, the paradigm itself will largely determine the range of admissable problems although we can expect external factors to mediate in the selection of problems which are considered to be of practical importance. The breakdown of normal science again comes from an interplay of internal factors (built up of unsolved puzzles) and external factors — thus Kuhn claims that the pressure for calendar reform was instrumental in a pre-Copernican crisis for Ptolemaic astronomy. At the stage of revolution we can, at least in principle, expect to observe a role for external factors in the formulation of the conceptual changes of the new paradigm. The rigorous reconstruction of concrete episodes of scientific progress in these terms has yet to be achieved, but the following readings will provide some insight, additional to those provided by the Galileo affair, into the role of external factors. It is not intended that this book should even attempt to exhaust this theme, and the readings suggested are limited to investigations into the period of the scientific revolution.

Readings:
(i) Basalla (1968) Introduction
Basalla's collection of papers (two of which are recommended below) is addressed to the specific question of the role of internal and external factors in the rise of modern science. The introduction is a comprehensive outline of this problem area.
(ii) Hessen (1931), *The Social and Economic Roots of Newton's Principia,* in Basalla (1968)
A Soviet Marxist philosopher provides a stimulating polemic account of Newton's work in which he attempts to demonstrate that its key themes were a reflection of the pervading technical problems of his time.
(iii) Merton (1938), *Science Technology and Society in Seventeenth Century England,* in Basalla (1968)
Merton argues that the 'Protestant ethic' contributed to the growth of modern science through a set of shared values. Also he demonstrates, in a manner similar to Hessen's but less extreme, the relation between the practical needs of the seventeenth century and the scientific researches of the period.

3(a) DOES A SCIENTIFIC APPROACH TELL US *ALL* THAT WE CAN KNOW ABOUT THE WORLD WE LIVE IN?

We can start to answer this question on the basis of the understanding of scientific knowledge derived from the above readings. We have gone beyond the view that we can assume the existence of a material reality which is a one to one reflection of the constructions of our science. The view which replaces it is of a world the details of which we cannot know but whose behavior we can tentatively and provisionally predict by means of mental models which we construct ourselves. What we do not know is whether aspects of the world exist which the mode of comprehension that we call scientific is not *able* to comprehend. To a certain extent this argument can be presented as a philosophical subterfuge which excuses the rejection of scientific rationality and encourages the adoption of metaphysical attitudes which are intellectually, and in political implication, chaotic. And yet its strength cannot be denied and we must try and consider its implications. The issue embraces sub-problem 3(b).

3(b) CAN A FORMAL DISTINCTION BE DRAWN BETWEEN HUMAN EXPERIENCES SUCH AS LOVE AND BEAUTY, AND HUMAN EXPERIENCES OF MASS, LENGTH AND TIME?

The implications of the line that we have been following would deny this distinction. Beauty and length are both mental categories into which we pack aspects of the experiences of our minds. Length can be measured and beauty cannot, but this can be interpreted as an indication of the limits of the scientific mode of analysis. Koestler (see below) quotes Bertrand Russell:

> Physics is mathematical not because we know so much about the physical world but because we know so little: it is only its mathematical properties that we can discover.

There are no easy answers to this problem and each student, each teacher, will have to make up his own mind. The readings below are all pertinent to the issue.

Readings:
(i) Koestler (1959) especially Epilogue
The 'divorce of faith and reason' which Koestler sees in the Galileo affair leads on to his discussion of the inadequacy of the scientific world-view, and the lack of moral and ethical progress which has accompanied the dramatic progress of science.

Koestler clearly recognizes that the scientific method cannot be applied to every aspect of human experience, and indeed he believes that 'our hypnotic enslavement to the numerical aspects of reality has dulled our perception of non-quantitative moral values'. Yet

Koestler cannot be dubbed an anti-rationalist; possibly the best label for Koestler is a liberal rationalist one. To Koestler, paranormal phenomena (e.g. extra-sensory perception) might lie outside the framework of our science, even of all science, and yet we cannot assume *a priori* that they do not exist. This particular subject is covered in Koestler's book *The Roots of Coincidence* (Hutchinson, 1972).

(ii) Roszak (1970) especially Chapter 7 and Appendix
(iii) Roszak (1973) especially parts 1 and 3
Roszak attacks scientific rationality head on. Scientific rationality to Roszak, sets the boundaries to what modern Western societies believe to be true and what they believe to be false, to what they believe to be subjective and what they believe to be objective, to what they understand as *knowledge*. Thus science defines our *reality principle*, the principle which bounds even our most revolutionary politics and, claims Roszak, dooms us.

This reality principle is what William Blake, and Roszak after him, calls *single vision* and represents a reduction of human consciousness. The sort of reduction which Blake, Wordsworth and Goethe cried out against.

Roszak can be usefully approached via his notion of scientific knowledge as a map (Roszak (1973) pp. 313 and 407) which takes a section through a rich and multidimensional world and thus reduces it to an abstraction which suits much that is real.

Roszak's work offers a fascinating and powerful indictment of science and of technocratic modes of thinking. Perhaps we might wonder whether technocracy *in fact* holds the power which Roszak credits it with, or whether human values (certainly derived from no objective consciousness) are able to triumph (sometimes in fact and always in principle) while decision making is, thankfully, conducted in the realm of politics and only rarely, and often unsuccessfully do technocratic methods intrude.

(iv) Easlea (1973) especially Chapter 10
Easlea's thesis can be seen in part as a refreshing corrective to Roszak. In order to build a beautiful world, capitalist society must be transcended. In this world, physics 'could well become a kind of delightful play, providing men with nourishment for their intellectual and aesthetic faculties'. But even Easlea is caught in a cleft stick in relation to the manipulative aspect of scientific knowledge. He advocates a life in harmony with nature but admits that since disease is ugly, medicine would come to be regarded as the noblest of the sciences. In conceding a value judgement against the malarial mosquito, Easlea makes harmony with nature a relative rather than an absolute imperative. Ultimately, Easlea's thesis differs little in its attitudes to science from that of less romantic thinkers: it is the political structure of the world, rather than its acceptance or

rejection of the 'scientific world-view' which will ultimately determine its degree of beauty.

A possible tutorial or seminar program

A set of five tutorial outlines is presented below; the outlines suggest further reading (often by reference to the previous section) which is strongly recommended as reading for the seminar leader.

A set of questions is suggested for each tutorial. In tutorials 1 to 4 it is intended that these should be posed directly and should form the basis of a discussion of the major problem area to which the tutorial relates. If used in the order in which they are presented below they will lead to a particular conclusion.

TUTORIAL 1: ON THE HISTORICAL MATERIAL

The intention of this tutorial is to consolidate understanding of the historical context in which Galileo's statements on science were made.

Relevant reading:
Main text, Chapters 1–8. Further reading from the Appendix and Bibliography Section G.

Questions:
What is meant by retrogression of the planets?
What is an epicycle?
How does it account for retrogression?
What other geometric devices were used by Ptolemy?
Was the Ptolemaic system accurate?
Was it scientific?
Was Copernicus' heliocentric hypothesis totally new?
What were its advantages in 1543?
Why was the preface to *De Revolutionibus* important?
Who wrote it?
What were the disadvantages of the Copernican system?
What were the characteristics of the Tychonic system?
What discoveries were reported by Galileo in *The Starry Messenger*?
Were these evidence for the Copernican theory?
What further discoveries did Galileo make with his telescope?
Were these evidence for the Copernican theory?
What was the attitude of the Church towards these discoveries?
What was Galileo's attitude? Was it scientifically sound?
Who was Foscarini?
Who was Bellarmine?
Why did Galileo write the *Letter to the Grand Duchess Christina*?
What steps were taken against Galileo in 1616?

Why did Galileo feel able to continue his work in 1623?
When was *Dialogue on the Great World Systems* published?
What were the consequences?

Essay or Discussion Questions:
If Earth were the only planet of the sun, might the history of astronomy have been different?

Why was the Ptolemaic system accepted for more than a thousand years?

Copernicus has been called a conservative. Is this fair?

Was the Church's treatment of Galileo anything less than scrupulously fair?

TUTORIAL 2: ON THE NATURE AND MEANING OF SCIENTIFIC KNOWLEDGE

The intention of this tutorial is to facilitate a discussion of issues 1(a)–(c) based firmly on the problems raised by Galileo's treatment of the Copernican theory.

Relevant reading:
Main text, Chapters 1–8. Further reading from the Appendix. Issues 1 (a)–(c) (see pp. 63–65). Bibliography Section A.

Questions:
Did Galileo believe that the Copernican theory represented absolute reality?
How does this differ from the belief that it merely 'saves the appearances'?
If we assume that the universe is in some way made of 'matter', i.e. reality is material, do we observe this 'matter' or do we just know the effects that it has on our minds?
Suppose that the earth was stationary, and that all the other heavenly bodies, including the stars, moved so as to present exactly the same appearance when viewed from the earth — would you be able to tell the difference? (Note to teacher — if the students are clever, it may be necessary either to argue that the Coriolis force, on the basis of the General Theory of Relativity, would still exist, or to suppose another material cause of the Coriolis force).
How, then, do we know that the material reality is not of this form?
If we don't know the ultimate material reality, how do we know that reality *is* material?
What are the alternative assumptions?
Did Galileo think that the Copernican theory was proven?
Was he right?

What would have constituted a proof?

Would this proof, if any, ensure that the theory would be true for all time?

What constituted a refutation of the Ptolemaic theory?

What if the Ptolemaic theory were right, and this apparent disproof were the result of some other unknown cause; would we be sure of knowing?

Can theories be *conclusively* proved or disproved?

What do we know about the nature of scientific theories? (Note to teacher — discussion based explicitly on Kuhn).

So does science guarantee certainty?

Does it rule out on *a priori* grounds the possibility of religious revelation as a source of knowledge?

Is it fair to say that Galileo was rational and the Church irrational?

TUTORIAL 3: ON THE EXTERNAL RELATIONS OF SCIENCE

The intention of this tutorial is to show that science is not neutral, does not exist in moral and social isolation, and is capable of reflecting even in the nature of its theories, socially derived concepts and assumptions.

Relevant reading:
Main text, especially edited readings. Further reading from the Appendix. Issues 1 (a)—(c) and 2 (a)—(b) (see pp. 63—67). Bibliography Sections A and B

Questions:
In what ways do you think science affects society?

Do you think that science can affect society in ways other than those associated with technology?

What do you think would have been the implications for the Catholic Church (and for seventeenth century Italy) of a total acceptance of the Copernican theory?

Did Galileo recognize the social implications of his science?

What do you imagine that his attitude to these implications might have been?

Did he act in a socially responsible fashion?

Did the involvement of the Church substantially affect the outcome between Ptolemaic and Copernican astronomy?

Why do you think the Ptolemaic system was changed after 1300 years? Did external factors play a part?

Do such factors merely act to slow or speed up the march of scientific progress, or could they actually influence the nature of the theories which are suggested and adopted?

Is it relevant to this point to know that Copernicus was influenced by the so-called Hermetic tradition which taught that the sun was a symbol of the Godhead? (Note to teacher — see for example Kearney (1971)).

Do you now believe that science is 'objective knowledge' which can be studied and understood without reference to wider social implications?

TUTORIAL 4: ON THE ADEQUACY OF THE SCIENTIFIC WORD-VIEW

The intention of this tutorial is to consider the implications of the conclusions reached in previous sections.

Relevant reading:
Previously suggested readings. Issues 3 (a)—(b) (see pp. 68—70). Bibliography especially Section C.

Questions:
From previous tutorials, we can deduce that scientific theories are intellectual constructs which we use to order our experience of the physical world?

What basic assumptions underlie the scientific word-view? (e.g. the mechanical analogy)

Is it right to assume that reality corresponds to this view?

What are the consequences of this assumption? (e.g. Laplacian determinism)

Are there human experiences which are not measurable or expressible in mathematics?

In the mechanical world view, how are these experiences accounted for?

Is science dehumanizing?

Some writers have argued that science *is* dehumanizing and have suggested that we reject it. Do you accept the consequences of this view?

TUTORIAL 5: A CLOSER EXAMINATION OF THE *LETTER TO THE GRAND DUCHESS CHRISTINA*

The intention of this tutorial is to examine, in the light of insights gained over previous sessions, the points that are made by Galileo in the *Letter to the Grand Duchess Christina*.

Relevant readings:
Main text, especially edited readings.

Questions:
The following questions are not intended to be used in a sequential fashion, as have been the questions for previous tutorials. Rather, they are intended as the basis for extended discussions for which students might be asked to prepare.

Alternatively they might be used for essay or for examination purposes.

To what extent does Galileo exhibit 'naive confidence in the power of the physical method' (Duhem)?

What is Geymonat's view of the implications of the Galilean distinction between scientific language and ordinary language? Do you share this view? Where these dangerous implications realized?

Why did Galileo believe that theology deserved the title 'queen of the sciences'? What was the significance of this view as opposed to the 'traditional theologians' view?

Was 'the shifting of the burden of proof' (Koestler) mere subterfuge?

Where does the burden of proof or disproof lie in modern science?

Assess Galileo's treatment of the miracle of Joshua.

Do you think that Galileo was anti-cleric?

Why do you think his campaign failed?

Why do you think he ignored Kepler's work?

Bibliography

Section G Historical and General

Brecht, B. (1963). *The Life of Galileo.* London, Methuen

Butterfield, H. (1968). *Origins of Modern Science.* London, G. Bell and Sons (first published 1949)
An excellent review of the period 1300–1800 written by a general historian.

de Santillana, G. and Stillman Drake (1959). 'Arthur Koestler and his Sleepwalkers' *Isis,* **50,** 3
A highly critical discussion of Koestler (op. cit.) by two of the foremost Galileo scholars. Accuses Koestler of a highly selective use of data in his unsympathetic portrayal of Galileo.

de Santillana, G. (1961). *The Crime of Galileo.* London, Mercury Books (first published in USA, 1955)
A scholarly examination of the relationship between Galileo and the Church authorities.

Galileo Galilei (1614–15). *Letter to the Grand Duchess Christina.* In Stillman Drake op. cit.

Geymonat, L. (1965). *Galileo Galilei.* New York, McGraw Hill

Hull, L.W.H. (1959). *History and Philosophy of Science.* London, Longmans
An excellent simplified general survey of the history of science and of the changing nature of scientific ideas.

Kearney, H. (1971). *Science and Change 1500–1700.* London, World University Library
An excellent introduction to the history of science in this period. Easy to read and beautifully illustrated, it is also an impressive analysis of the rise of the mechanist tradition in science.

Koestler, A. (1959). *The Sleepwalkers.* London, Hutchinson (also Penguin Books, 1964 and 1968)

Kuhn, T.S. (1957). *The Copernican Revolution.* Cambridge, Mass., Harvard University Press (also in paperback: New York, Vintage Books, 1959)

Stillman Drake (1957). *Discoveries and Opinions of Galileo.* New York, Doubleday

Note that these texts constitute a highly selective sample. Galileo is perhaps the most widely written about of all scientific figures, and an immense amount of literature exists. *Those wishing to pursue the historical materials further than the above list allows will find excellent bibliographies in many of the listed books.* Kearney can be particularly recommended for an intermediate level bibliography.

Section A   Instrumentalism and reality

Duhem, P. (1954). *The Aim and Structure of Physical Theory.* Princeton, N.J., Princeton University Press (first published in France, 1906)

Galileo Galilei. *The Assayer* reprinted in Stillman Drake (op. cit.) See especially Galileo's distinction between subjective and objective qualities (p. 274 in Doubleday edition) and compare this with John Locke's distinction between primary and secondary qualities. A world in which only shape, number and motion are 'real' is the world of the mechanical paradigm.

Hull (op. cit) contains an excellent introduction to the epistemological ideas of Plato, Locke, Berkeley and Hume.

Kuhn, T.S. (1962). *The Structure of Scientific Revolutions.* Chicago, Ill., University of Chicago Press (second edition 1970)

Popkin *et al.* (1969). *Philosophy Made Simple.* London, W.H. Allen Chapter five is a particularly clear and concise summary of epistemological problems in the historical context. Plato to Hume. Very useful background, particularly when read in conjunction with Hull, to the problems of 'reality'.

Popper, K. (1963). *Conjectures and Refutations.* London, Routledge and Kegan Paul

Russell, B. (1968). *A History of Western Philosophy.* London, Allen & Unwin
A useful source of information on individual philosophers such as Locke, Berkeley and Hume.

Section B   Proof and disproof

Kuhn, T.S. (op. cit., as Section A)

Lakatos and Musgrave (eds.) (1970). *Criticism and the Growth of Knowledge.* London, Cambridge University Press
A collection of essays on the Kuhn—Popper debate. Lakatos' own contribution is an interesting development of the Popperian position and can act as a useful source of insight into Popper's works at a level more advanced than that represented in the chapters of Popper (1963) recommended in this text.

Magee, B. (1973). *Popper.* London, Fontana Modern Masters
An interesting commentary on Popper's philosophy of science which demonstrates the relationship between that and his political philosophy.

Medawar, P.B. (1969). *Induction and Intuition in Scientific Thought.* London, Methuen
An interesting simple account by an eminent scientist of the problems of scientific methodology. Can be confusing if not read

critically — is Medawar offering a description of scientific methodology or an epistemological appraisal? The answer often seems to be unclear even in the writer's mind. Advocates a Popperian viewpoint but does not attend to the problems raised by the Kuhnian critique.

Popper (op. cit., as Section A)

Section C The external relations of science

Basalla (ed.) (1968). *The Rise of Modern Science.* Lexington, Mass. Heath

Hessen (1931). *The Social and Economical Roots of Newton's Principia.* In Basalla (op. cit.)

Jevons, F.R. (1973). *Science Observed.* London, Allen & Unwin

Merton (1938). *Science Technology and Society in Seventeenth Century England.* In Basalla (op. cit.)

Ravetz, J. (1971). *Scientific Knowledge and its Social Problems.* Oxford, Clarendon Press (also Penguin Education, 1973)
Read as a whole, the book contributes to the understanding of scientific knowledge as a socially regulated construction. Germane to present discussion of external effects is Chapter 2 which discusses the social problems of industrialized science.

Sklair, L. (1973). *Organised Knowledge.* London, Paladin, Hart-Davis
As a whole, the book falls into the same category as Ravetz (op. cit.) constituting an examination of science from a post-Kuhnian sociological/epistemological point of view. But see particularly Chapter 6 on the social functions of science.

Section D The implications and limits of the scientific world-view

Barnes, B. (ed.) (1972). *Sociology of Science.* Harmondsworth, Penguin Modern Sociology Readings
Part six, especially Chapter 19 for a taste of Marcuse.

Easlea, B. (1973). *Liberation and the Aims of Science.* London, Chatto and Windus

Koestler, A. (1959). op. cit. See p. 68

Koestler, A. (1972). *The Roots of Coincidence.* London, Hutchinson

Mays, W. (1973). *Koestler,* Makers of Modern Thought. Woking, Lutterworth Press

Roszak, T. (1970). *The Making of a Counterculture.* London, Faber and Faber

Roszak, T. (1973). *Where the Wasteland Ends.* London, Faber and Faber (See p. 69)

Appendix:
Further Reading

1. Arthur Koestler (1959). *The Sleepwalkers.* London, Hutchinson.
Also in paperback: Harmondsworth, Penguin Books, 1964 and 1968.

An extended and novelish account which broadly covers the material
contained in this book and stresses the relevance of the personal
lives of Copernicus, Kepler and Galileo to their scientific achievements. Science to Koestler is more of a 'sleep walking' performance
than the product of logical enquiry, and this aspect of Kepler's
work in particular is emphasized.

Koestler is a critic of science which poses such questions as 'is
man a machine?'; for are not human emotions, is not human free-will,
as basic a set of experiences on which the edifice of scientific knowledge is constructed? Thus Koestler is critical of the divorce of
science and religion, of so-called faith and reason. By distinguishing
between the physical world and the spiritual world we have, Koestler
suggests, brought about a unique increase in man's physical power
and a decline in his moral or spiritual awareness. Koestler seems to
blame Galileo for this state of affairs.

Whether we agree or disagree with Koestler's analysis we cannot
but be stimulated by it. Once this book has been read, a useful entry
point to *The Sleepwalkers* is the 'Epilogue' in which Koestler discusses many of the points outlined above. The full text could usefully
be read. Otherwise, it is well indexed, and can serve as a reference
to points of detail which might not be covered to the reader's
satisfaction in this book.

Finally, anybody reading *The Sleepwalkers* will become aware of
the extent to which the author of this book is indebted to Koestler's
account and Koestler's ideas.

2. Thomas S. Kuhn (1957). *The Copernican Revolution.* Cambridge,
Mass., Harvard University Press. Also in paperback: New York,
Vintage Books, 1959

Kuhn provides a detailed account of astronomy up to and including
the Copernican revolution. His account dwells on the technical details
of the various astronomical systems that he discusses, and these
details are explained clearly and concisely.

Kuhn's concern is with the nature of the change from a Ptolemaic
to a Copernican astronomy, and many of the ideas later to be incorporated in his seminal work *The Structure of Scientific Revolutions*
can be read in his account of the Copernican Revolution. He is less

concerned with Galileo; more with the nature of the new science that the Copernican revolution precipitated.

Some students find this account of the two-sphere universe and of Copernicus' innovation preferable to that of Koestler. Chapter 4 which discusses the assimilation of Aristotelian Ptolemaic cosmology over the first 1500 years AD is particularly interesting and valuable.

3. L. Geymonat (1965). *Galileo Galilei.* New York, McGraw Hill (first published in Italian 1957)

Where Koestler emphasizes the negative aspect of Galileo's achievements, Geymonat concentrates on the positive. He presents Galileo's battle with the Church as a propaganda campaign designed to secure for science the sustenance and aid that it needed from all who occupied positions of power in society. To Geymonat, Galileo was not concerned with religious problems but with the promotion of science as a matter of public interest.

Geymonat focuses on the methodological and philosophical aspects of Galileo's conception of science, his conception of scientific facts and his alliance of mathematical and logical reasoning (for long the preserve of those who would interpret nature by arguing from first principles) with sense experience, experiment and observation.

To a certain extent, then, Geymonat's analysis goes beyond the arguments presented in this book. But while his methodological discussions are at a relatively high level of academic abstraction, they are clearly and succinctly expressed. The book is, moreover, rich in historical detail and worthy of serious examination.

4. Stillman Drake (translator) (1957). *Discoveries and Opinions of Galileo.* New York, Doubleday

Drake has translated four of Galileo's works, *The Starry Messenger, Letters on Sunspots, Letter to the Grand Duchess Christina,* and *The Assayer* and presents them linked together by a lengthy original text. The text is an interesting additional source of historical information and comment, but the main interest in the book lies in the works of Galileo.

It is impossible to read *The Starry Messenger* or *The Assayer* without being impressed by the lucidity of Galileo's style and the modernity of his mode of scientific reasoning. Some perusal of these texts is a pre-requisite for an understanding of the mind of Galileo.

5. Bertolt Brecht (1963). *The Life of Galileo.* London, Methuen

This is a dramatic account of the Galileo affair written by a Marxist playwright; the final version of the play was completed soon after

the dropping of the atomic bombs on Japan and during the conse-
quential subjugation of scientific research to the political dictates of
the Cold War. Mistrust of science now led Brecht to claim that 'It
had become a disgrace to discover anything'.

Brecht's attitude to Galileo is not dissimilar to Koestler's. He
writes 'The fact is that Galileo enriched astronomy and physics by
simultaneously robbing these sciences of a greater part of their social
importance' and concludes 'The atom bomb is, both as a technical
and as a social phenomenon, the classical end-product of his contribution to science and his failure to society'.

It is possible to discover in the play a number of statements on
the relations between science and society. And yet the play was
originally written before Hiroshima and can also be interpreted less
as a play about science and more as a play about radical change.
About the commitment to radical change in the face of reaction and
on a basis of faith rather than in the light of compelling and conclusive evidence. About Brecht's own faith?

In this book we expect to use Brecht's play as no more than a
vehicle for further insight into Galileo and the nature of his campaign. The question 'What is the play about?' merits serious consideration but must be exterior to this book. Some light is thrown on
both aspects of the play by two articles in the journal *Technology
and Society*: F.R. Jevon's 'Brecht's Life of Galileo and the Social
Relations of Science', *Technology and Society*, vol. 4, 1968 and M.A.
Cohen, 'Brecht's Galilei: a continuing discussion', *ibid*, vol. 5.